Introduction to Biomedical Engineering: Biomechanics and Bioelectricity Part I

Synthesis Lectures on Biomedical Engineering

Editor
John D. Enderle, *University of Connecticut*

Introduction to Biomedical Engineering: Biomechanics and Bioelectricity - Part I
Douglas A. Christensen
2009

Basic Feedback Controls in Biomedicine
Charles S. Lessard
2008

Understanding Atrial Fibrillation: The Signal Processing Contribution, Volume II
Luca Mainardi, Leif Sörnmon, and Sergio Cerutti
2008

Understanding Atrial Fibrillation: The Signal Processing Contribution, Volume I
Luca Mainardi, Leif Sörnmon, and Sergio Cerutti
2008

Introductory Medical Imaging
A. A. Bharath
2008

Lung Sounds: An Advanced Signal Processing Perspective
Leontios J. Hadjileontiadis
2008

An Outline of Information Genetics
Gérard Battail
2008

Neural Interfacing: Forging the Human-Machine Connection
Thomas D. Coates, Jr.
2008

iv

Quantitative Neurophysiology
Joseph V. Tranquillo
2008

Tremor: From Pathogenesis to Treatment
Giuliana Grimaldi and Mario Manto
2008

Introduction to Continuum Biomechanics
Kyriacos A. Athanasiou and Roman M. Natoli
2008

The Effects of Hypergravity and Microgravity on Biomedical Experiments
Thais Russomano, Gustavo Dalmarco, and Felipe Prehn Falcão
2008

A Biosystems Approach to Industrial Patient Monitoring and Diagnostic Devices
Gail Baura
2008

Multimodal Imaging in Neurology: Special Focus on MRI Applications and MEG
Hans-Peter Müller and Jan Kassubek
2007

Estimation of Cortical Connectivity in Humans: Advanced Signal Processing Techniques
Laura Astolfi and Fabio Babiloni
2007

Brain-Machine Interface Engineering
Justin C. Sanchez and José C. Principe
2007

Introduction to Statistics for Biomedical Engineers
Kristina M. Ropella
2007

Capstone Design Courses: Producing Industry-Ready Biomedical Engineers
Jay R. Goldberg
2007

BioNanotechnology
Elisabeth S. Papazoglou and Aravind Parthasarathy
2007

Bioinstrumentation
John D. Enderle
2006

Fundamentals of Respiratory Sounds and Analysis
Zahra Moussavi
2006

Advanced Probability Theory for Biomedical Engineers
John D. Enderle, David C. Farden, and Daniel J. Krause
2006

Intermediate Probability Theory for Biomedical Engineers
John D. Enderle, David C. Farden, and Daniel J. Krause
2006

Basic Probability Theory for Biomedical Engineers
John D. Enderle, David C. Farden, and Daniel J. Krause
2006

Sensory Organ Replacement and Repair
Gerald E. Miller
2006

Artificial Organs
Gerald E. Miller
2006

Signal Processing of Random Physiological Signals
Charles S. Lessard
2006

Image and Signal Processing for Networked E-Health Applications
Ilias G. Maglogiannis, Kostas Karpouzis, and Manolis Wallace
2006

© Springer Nature Switzerland AG 2022
Reprint of original edition © Morgan & Claypool 2009

All rights reserved. No part of this publication may be reproduced, stored in a retrieval system, or transmitted in any form or by any means—electronic, mechanical, photocopy, recording, or any other except for brief quotations in printed reviews, without the prior permission of the publisher.

Introduction to Biomedical Engineering: Biomechanics and Bioelectricity - Part I

Douglas A. Christensen

ISBN: 978-3-031-00508-4   paperback
ISBN: 978-3-031-01636-3   ebook

DOI 10.1007/978-3-031-01636-3

A Publication in the Springer series
SYNTHESIS LECTURES ON BIOMEDICAL ENGINEERING

Lecture #28
Series Editor: John D. Enderle, University of Connecticut

Series ISSN
Synthesis Lectures on Biomedical Engineering
Print 1932-0328   Electronic 1932-0336

Introduction to Biomedical Engineering: Biomechanics and Bioelectricity Part I

Douglas A. Christensen
University of Utah

SYNTHESIS LECTURES ON BIOMEDICAL ENGINEERING #28

ABSTRACT

Intended as an introduction to the field of biomedical engineering, this book covers the topics of biomechanics (Part I) and bioelectricity (Part II). Each chapter emphasizes a fundamental principle or law, such as Darcy's Law, Poiseuille's Law, Hooke's Law, Starling's law, levers and work in the area of fluid, solid, and cardiovascular biomechanics. In addition, electrical laws and analysis tools are introduced, including Ohm's Law, Kirchhoff's Laws, Coulomb's Law, capacitors and the fluid/electrical analogy. Culminating the electrical portion are chapters covering Nernst and membrane potentials and Fourier transforms. Examples are solved throughout the book and problems with answers are given at the end of each chapter. A semester-long Major Project that models the human systemic cardiovascular system, utilizing both a Matlab numerical simulation and an electrical analog circuit, ties many of the book's concepts together.

KEYWORDS

biomedical engineering, biomechanics, cardiovascular, bioelectricity, modeling, Matlab

To Laraine

Contents

Synthesis Lectures on Biomedical Engineering iii

Contents xi

Preface xv

1   Basic Concepts: Numbers, Units and Consistency Checks 1

   1.1   Introduction 1

   1.2   Numbers and significant figures 1

       1.2.1 Scientific Notation   2

       1.2.2 Accuracy and Precision   3

       1.2.3 Significant Figures in Calculations   3

   1.3   Dimensions and units 5

       1.3.1 SI Units   6

       1.3.2 Keeping Track of Units in Equations   8

       1.3.3 English and Other Units   8

   1.4   Conversion factors 8

       1.4.1 The Use of Weight to Describe Mass   10

   1.5   Consistency checks 10

       1.5.1 Reality Check   11

       1.5.2 Units Check   11

       1.5.3 Ranging Check   12

   1.6   Organization of the remaining chapters 12

   1.7   Problems 13

2   Darcy's Law: Pressure-Driven Transport Through Membranes 15

   2.1   Introduction – Biological and Man-Made Membranes 15

       2.1.1 Man-Made Membranes   17

2.2   Darcy's Law 18

    2.2.1   Ideal and Nonideal Materials   21

2.3   Mechanical Filtration (Sieving) 22

2.4   Problems 25

3   Poiseuille's Law: Pressure-Driven Flow Through Tubes 27

3.1   Introduction – Biological Transport 27

3.2   Poiseuille's Law 30

    3.2.1   Simplified Version of Poiseuille's Law   33

    3.2.2   Assumptions for Poiseuille's Law   34

3.3   Power Expended in the Flow 36

3.4   Series and Parallel Combinations of Resistive Elements 36

    3.4.1   Series   37

    3.4.2   Parallel   37

3.5   Problems 42

4   Hooke's Law: Elasticity of Tissues and Compliant Vessels 45

4.1   Introduction 45

4.2   The Action of Forces to Deform Tissue 45

4.3   HOOKE'S LAW AND ELASTIC TISSUES 46

4.4   Compliant Vessels 50

4.5   Incompressible Flow of Compliant Vessels 53

4.6   Problems 55

5   Starling's Law of the Heart, Windkessel Elements and Conservation of Volume 61

5.1   Introduction – Compliance of the
    Ventricles 61

5.2   Pressure-Volume Plots: the pv Loop 62

5.3   STARLING'S LAW OF THE HEART 64

5.4   Windkessel Elements 67

5.5   Conservation of Volume in Incompressible Fluids 68

5.6   Problems 70

6   Euler's Method and First-Order Time Constants 73

    6.1   Introduction – Differential Equations 73

    6.2   Euler's Method.. 74

    6.3   Waveforms of Pressure and Volume 75

    6.4   First-Order Time Constants.. 76

    6.5   Problems 84

7   Muscle, Leverage, Work, Energy and Power...................................... 87

    7.1   Introduction – Muscle 87

    7.2   Levers and Moments 87

    7.3   Work.. 92

    7.4   Energy 92

    7.5   Power 94

          7.5.1   Power in Fluid Flow   94

    7.6   Problems 95

A   Conversion Factors 97

B   Material Constants 99

    B.1   Viscosity.. 99

    B.2   Density and Specific Gravity 99

    B.3   Permeability 100

    B.4   Young's Modulus and Ultimate Stress 100

    Bibliography 101

Preface

NOTE ON ORGANIZATION OF THIS BOOK

The material in this book naturally divides into two parts:

1. Chapters 1-7 cover fundamental biomechanics laws, including fluid, cardiovascular, and solid topics (1/2 semester).

2. Chapters 8-15 cover bioelectricity concepts, including circuit analysis, cell potentials, and Fourier topics (1/2 semester).

A Major Project accompanies the book to provide laboratory experience. It also can be divided into two parts, each corresponding to the respective two parts of the book. For a full-semester course, both parts of the book are covered and both parts of the Major Project are combined.

The chapters in this book are support material for an introductory class in biomedical engineering[1] They are intended to cover basic biomechanical and bioelectrical concepts in the field of bioengineering. Coverage of other areas in bioengineering, such as biochemistry, biomaterials and genetics, is left to a companion course. The chapters in this book are organized around several fundamental laws and principles underlying the biomechanical and bioelectrical foundations of bioengineering. Each chapter generally begins with a motivational introduction, and then the relevant principle or law is described followed by some examples of its use. Each chapter takes about one week to cover in a semester-long course; homework is normally given in weekly assignments coordinated with the lectures.

The level of this material is aimed at first-semester university students with good high-school preparation in math, physics and chemistry, but with little coursework experience beyond high school. Therefore, the depth of explanation and sophistication of the mathematics in these chapters is, of necessity, limited to that appropriate for entering freshman. Calculus is not required (though it is a class often taken concurrently); where needed, finite-difference forms of the time- and space-varying functions are used. Deeper and broader coverage is expected to be given in later classes dealing with many of the same topics.

Matlab is used as a computational aid in some of the examples in this book. Where used, it is assumed that the student has had some introduction to Matlab either from another source or from a couple of lectures in this class. In the first half of the cardiovascular Major Project discussed below,

---

[1] At the University of Utah, this course is entitled Bioen 1101, Fundamentals of Bioengineering I.

Matlab is used extensively; therefore, the specific Matlab commands needed for this Major Project must be covered in class or in the lab if this particular part of the project is implemented.

A Major Project accompanies these chapters at the end of the booklet. The purpose of the Major Project, a semester-long comprehensive lab project, is to tie the various laws and principles together and to illustrate their application to a real-world bioengineering/physiology situation. The Major Project models the human systemic cardiovascular system. The first part of the problem takes approximately one-half of a semester to complete; it uses Matlab for computer modeling the flow and pressure waveforms around the systemic circulation. Finite-difference forms of the flow/pressure relationships for a lumped-element model are combined with conservation of flow equations, which are then iterated over successive cardiac cycles. The second half of the problem engages a physical electrical circuit to analyze the same lumped-element model and exploits the duality of fluid/electrical quantities to obtain similar waveforms to the first part. This Major Project covers about 80% of the topics from the chapter lectures; the lectures are given "just-in-time" before the usage of the concepts in the Major Project.

Although the Major Project included with this book deals with the cardiovascular system, other Major Project topics may be conceived and substituted instead. Examples include modeling human respiratory mechanics, the auditory system, human gait or balance, or action potentials in nerve cells. These projects could be either full- or half-semester assignments.

ACKNOWLEDGEMENTS

The overarching organizational framework of these chapters around fundamental laws and principles was conceived and encouraged by Richard Rabbitt of the Bioengineering Department at the University of Utah. Dr. Rabbitt also provided much of the background material and organization of Chapters 2 and 4. Angela Yamauchi provided the organization and concepts for Chapter 3. David Warren contributed to the initial organization of Chapter 8. Their input and help was vital to the completion of this booklet.

Douglas A. Christensen
University of Utah
March 2009

CHAPTER 1

Basic Concepts: Numbers, Units and Consistency Checks

1.1 INTRODUCTION

Welcome to biomedical engineering, a very rewarding field of study! Biomedical engineering is the application of engineering principles and tools for solving problems in health care and medicine. Of all the engineering specialties, it is arguably the most interdisciplinary, requiring the study of biology, physiology and organic chemistry in addition to mathematics, physics and engineering topics. This makes the field particularly challenging as well as engrossing. Many professionals in biomedical engineering have chosen this field because they strive to improve the lives of their fellow humans and society.

The goal of this book is to introduce you to some basic concepts of biomedical engineering by covering several fundamental physical laws and principles that underlie biomedical engineering. This book is focused on the biomechanical and bioinstrumentation (electrical) aspects of the field. An optional Major Project dealing with modeling the human cardiovascular system ties many of the book's topics together. It is left to other texts to cover other important areas in the field, such as biochemical, molecular and biomaterial topics.

We start our studies with some basic concepts involving numbers and calculations.

1.2 NUMBERS AND SIGNIFICANT FIGURES

Numbers and values are the stock-in-trade of the engineering profession. Engineers are often involved in various measurements (for example, a bioengineer may design a device to measure the level of blood glucose in a new and novel way), and one of the distinguishing features of a successful engineer is that s/he strives to be quantitatively correct. This requires careful attention to the manipulation and display of numerical values.

An important characteristic of any numerical value is the number of digits, or **significant figures**, it contains. A significant figure is defined as *any digit in the number ignoring leading zeros and the decimal point*. For example:

821	has 3 significant figures
160.6	has 4 significant figures
160.60	has 5 significant figures
0.0310	has 3 significant figures
1.5×10^3	has 2 significant figures

How many significant figures should you use when writing a number? A commonly accepted rule in specifying a number is *to use as many, but not more, significant figures as can be reasonably trusted to be accurate in value.* For example, let's say you measure a tall classmate's height with a meter-long ruler. Which one of the following ways do you think has the appropriate number of significant figures to report the result of the height measurement?

    a. 2.0131 m   (with 5 significant figures)
    b. 2.01 m   (with 3 significant figures)
    c. 2 m   (with 1 significant figure)

Answer **a** is unrealistic, because it gives the impression that your classmate's height is measured with a precision or accuracy[1] of one-tenth of a millimeter, or 0.0001 m—about the thickness of a human hair, which is obviously not justified[2] since it depends upon the posture of the classmate at the time of the measurement, how much his hair was pushed down, how well the ruler was aligned with his spine, etc. Besides, it's impossible to read a meter ruler to this fine an increment anyway.

Answer **c** is also unreasonable for the opposite reason: it doesn't convey enough information about the accuracy of the measurement you took. It could describe anyone with a height between 1.50 m (even shorter than I am) and 2.49 m (taller than any NBA player). Surely you measured within a tighter range than that.

Answer **b** is the most reasonable choice. It tells the reader that in your best judgement, your measurement can be trusted to within about a centimeter, which is consistent with the uncertainties of the measurement. Therefore you report a value to the nearest 0.01 m (which is the **2.01** m answer), with three significant figures.

But what if your classmate measured exactly 2 m (within a fraction of a centimeter)? Do you report answer **c**? No. Your number should make it clear that you have confidence to the centimeter level, so in this case you should report a measurement of **2.00** m, with three significant figures.

1.2.1   SCIENTIFIC NOTATION

In light of the preceding discussion, numbers you come across which are multiples of 10 need to be interpreted with care. Consider the number 200. It is not clear whether the trailing zeros in this number show the confidence level of a measurement, or are simply decimal place holders. For instance, the length of the 200-m dash, a track-and-field event, is actually specified by the sport's rules to a high accuracy—to within 1 cm—so it really is the 200.00-m dash. On the other hand, a "200-m high cloud" only means a cloud whose distance off the ground is about 200 m, give-or-take approximately 10 m.

So, to avoid confusion and to be more specific about the intended precision, scientific notation is preferred by engineers in these situations. Using scientific notation, the examples in the previous

---

[1]The concepts of accuracy and precision are discussed later in this chapter.
[2]Whether any accuracy is justified or not is object dependent. The diameter of a small metal stent whose purpose is to keep open a coronary artery can (and should) be specified to one-tenth of a millimeter.

paragraph would be written as **2.0000 × 10^2** m and **2.0 × 10^2** m, respectively, indicating the intended precision.

1.2.2   ACCURACY AND PRECISION

These terms are often used interchangeably, but they really describe different things. **Accuracy** is a measure of how close a value is to the "true" value (as determined by some means). **Precision** is an indication of the repeatability of the measurement, if done again and again under the same circumstances. Thus a golfer would rather be accurate than precise if she repeatedly hooks her drives into the trees.

Accuracy and precision are sometimes specified in terms of a ± range indicating the uncertainties of a measurement, such as 56 ±3 mmHg or 18.5 ±0.8% of full scale. If unspecified, the uncertainty is assumed to be approximately one unit of the last significant digit of the number, as explained above.

1.2.3   SIGNIFICANT FIGURES IN CALCULATIONS

Modern computers allow the manipulation and display of numbers with many significant figures (up to 10 or more). But you must be careful not to mislead your reader when reporting your final answer. Use only the number of significant figures justified by the reasoning above. Here are some generally accepted rules that are consistent with that reasoning:

Addition and Subtraction - After the operations of addition and subtraction, *the final answer should contain digits only as far to the right as the rightmost decimal column found in the* least *precise number used in the calculation*. For example, consider the following:

$$\begin{array}{r} 31.5 \\ + \ 2.8925 \\ \hline 34.3925 \end{array}$$ (as displayed on your calculator)

Following the rule, you should round off [3] this answer to **34.4** before you report it. This is because the first number, 31.5, is the least precise and has significant digits on the right out to the 0.1 decimal column, so your final answer should be rounded to this same column.

Here's a thought problem explaining why: Suppose you were asked the value of your personal assets for an application for a scholarship. You own a computer (estimated value of $730, accurate to within about $10), clothes (estimated value of $23—you're a sharp dresser), and a bank account (balance of $63.21). Therefore you sum your total assets:

---

[3] Round off as follows: Round up the last digit kept in the answer by 1 if the next digit (the first one to be dropped) is 5 or greater; otherwise don't change the last kept digit. Usually the displays of calculators do this rounding automatically when set to the appropriate number of significant figures.

$730
+ 23
+ 63.21
$816.21   (as displayed on your calculator)

You might be tempted to put down $816.21 on the application form, but with a little thought that would be misleading. It would imply that you know the worth of your computer and clothes to the penny. More realistically, the implied accuracy of the answer should be no better than the least accurate of the parts (i.e., the estimated value of your computer). Using the rule above, **$820** is the best answer. (Actually, 8.2×10^2 is an even better answer, because it clearly shows that you're confident only to the $10 level).

More Examples:

0.1035
+ 0.0076
0.1111

The answer should be reported as **0.1111**.

16.732
- 0.11
16.622

The answer should be reported as **16.62**.

5.5
+ 17.83
+ 2.11
25.44

The answer should be reported as **25.4**.

2.0
- 0.0006
1.9994

The answer [4] should be reported as **2.0**.

Multiplication and Division – After the operations of multiplication and division, *the answer should contain only as many significant figures as found in the number with the* fewest *significant figures.*

---

[4]This rounded-off answer may appear strange since it appears as though no subtraction took place, but remember that since the original 2.0 number is assumed precise only to one decimal place, the answer can be no more precise than to one decimal place. (If, however, the original 2.0 number is really known to four decimal places, it would have been written as 2.0000. Then the final answer would be 1.9994.)

For example, when you multiply (83.6) x (10,858) the calculator display may show 907,728.8 but the final answer should be rounded to *three* significant figures: **908,000** (or even better, 9.08×10^5). As a division example, 563 / 6.2 = 90.81 should be reported as **91**.

Here's a thought problem to help explain why: Suppose you are given the task of finding the weight of one full Pepsi can. You decide to weigh a six-pack of Pepsi cans on a bathroom scale, and divide by 6 to get an "average" answer—not a bad idea. On the scale, you measure the weight of the six-pack (without any packaging material) at 4.7 lbf. (Note that you have recorded a reasonable number of significant figures—two—for this measurement since the bathroom scale you used can be read only to about the nearest 0.1 lbf.) But when you use your calculator to divide 4.7 lbf by 6, you get 0.78333 lbf as the weight of a single can. Obviously you can't report the weight of a Pepsi can to a precision of 0.00001 lbf (the weight of a grain of sand) when you originally measured the total six-pack to a precision of only 0.1 lbf. It is more logical to use the degree of precision of the least precise number in the division, which in this problem is *two* significant figures. So the final answer should be reported as **0.78** lbf. (But what about the divisor, the number 6? Doesn't it have only *one* significant figure? No; in this case, it is an *exact* number and therefore can be considered to have an infinite number of zeroes after the decimal point—i.e., an infinite number of significant figures. Thus the number with the fewest significant figures is the number 4.7.)

More Examples:

> (0.431)(0.002) = 0.000862 should be reported as **0.001**.
> 163.4 / 16.555 = 9.87013 should be reported as **9.870**.
> $(4.3 \times 10^8) / (6.241 \times 10^{-3}) = 6.890 \times 10^{10}$ should be reported as $\mathbf{6.9 \times 10^{10}}$.

More Complex Operations – When an operation of multiplying or dividing is combined with adding or subtracting to achieve the final answer, or when nonlinear operations such as logarithmic or trigonometric operations are performed, it's harder to determine the correct number of significant figures in the answer. Intermediate answers must be rounded at each step, which is often difficult and awkward to implement. Therefore, there is no simple rule to apply here, and you should apply commonsense using the spirit of the previous rules as much as possible.

1.3   DIMENSIONS AND UNITS

Specifying **dimensions** is a scheme for grouping and labeling similar physical quantities. Thus the letter L represents the dimension of any length, M the dimension of the mass of any body, T the dimension of time, etc. Dimensions are not the same as **units**, since there are several possible alternative units for measuring any dimension. For example, the length (of dimension L) of an object can be measured in units of meters, inches, yards, chains, miles, or even light-years.

There are two major systems of units in use in the world today:

- A metric system known as the Système International d'Unites, or **SI**, which is the international standard used in most scientific work, and

- A non-metric system called the **English** system, which is in widespread use in the United States and Britain, especially for non-scientific applications.

Engineers need to be familiar with both major systems of units and be able to readily convert any value back and forth between the two systems. Biomedical engineers in particular must be agile with unit conversions because they must deal with units that are traditional in clinical settings but which are outside even the two major systems, such as blood pressure measured in mmHg and cardiac output measured in L/min.

Any number reported without units is meaningless. For example, if your systolic blood pressure is 2.3 (with no units), are you in good shape or bad shape? Without specifying the units of measurement, it's impossible to tell. If in units of psi, the pressure is fine; if in units of mmHg, it's not good at all.

The importance of specifying units (and also of close communication within an engineering team) was painfully demonstrated by the following incident:

On September 23, 1999, contact was abruptly lost with the NASA Mars Climate Orbiter. It was later determined that due to an information error on earth, the spacecraft failed to enter a proper orbit around Mars and was lost. An investigation uncovered the cause: Two teams, one in California and one in Colorado, were responsible for coordinating the correct maneuvering of the spacecraft as it neared Mars. As unbelievable as it sounds, one team used English units (e.g., inches, feet and pounds) while the other used metric units (centimeters, newtons), and each group was unaware that the other was using different units! NASA has since taken steps to make sure that this type of error doesn't happen again.

1.3.1   SI UNITS

As mentioned, metric SI units are used throughout the world. The SI is an outgrowth of the older MKS system, where MKS stands for meter, kilogram and second (the units for length, mass and time respectively). These units are retained as three of the base units of the SI. The other base units are the ampere, mole, kelvin, and candela. In addition, various combinations of these base units can define many other derived units defined for often-used quantities. Table 1.1 lists the base units and some important derived units in the SI.

Prefixes – Sometimes the size of a unit system does not match well the scale of the quantity being measured. For example, a meter is okay as it stands to use for measuring the length of a building, but the diameter of a virus is much smaller than a meter, and the diameter of the earth is much larger. So standard multiplying factors represented by prefixes are used to scale the meter (or any other unit) to better match the measurement. Some common prefixes to scale the root unit by multiples of ten are given in Table 1.2.

For example, the diameter of a flu virus is approximately $d = 0.00000010$ m $= 1.0 \times 10^{-7}$ m $= 0.10$ μm. Notice how much more convenient the prefix notation is when specifying this value. Also note that the proper use of prefixes, like scientific notation, can avoid the ambiguity of how many significant figures are intended in a number that is a multiple of 10.

Table 1.1: Some Base and Derived SI Units			
Quantity	**Unit**	**Symbol**	**Equivalent to**
length	meter	m	(base unit)
mass	kilogram	kg	(base unit)
time	second	s	(base unit)
electrical current	ampere	A	(base unit)
material amount	mole	mol	(base unit)
temperature	kelvin	K	(base unit)
light intensity	candela	cd	(base unit)
force	newton	N	$kg{\cdot}m/s^2$
pressure	pascal	Pa	$N/m^2 = kg/(m{\cdot}s^2)$
energy	joule	J	$N{\cdot}m = kg{\cdot}m^2/s^2$
power	watt	W	$J/s = kg{\cdot}m^2/s^3$
frequency	hertz	Hz	$1/s$

Table 1.2: Common Prefixes		
Prefix	**Symbol**	**Multiplier**
femto–	f	10^{-15}
pico–	p	10^{-12}
nano–	n	10^{-9}
micro–	μ	10^{-6}
milli–	m	10^{-3}
centi–	c	10^{-2}
kilo–	k	10^{3}
mega–	M	10^{6}
giga–	G	10^{9}
tera–	T	10^{12}

Rules for the Use of SI Units – When writing SI units, some generally accepted rules apply. These rules are intended to improve the consistency of SI unit usage and reduce the chance of misinterpretation.

A good summary of these rules can be found at the National Institute of Standards and Technology (NIST) website (`http://physics.nist.gov/cuu/Units/checklist.html`).

1.3.2   KEEPING TRACK OF UNITS IN EQUATIONS

It is very helpful, whenever possible, to carry units along in the numerical solution of equations. For example, consider the problem of finding the force needed to accelerate a 5.0 kg mass to an acceleration of 1.6 cm/s^2. The physical law that applies here is Newton's second law:

$$F = ma \tag{1.1}$$

where F is the value of the force, m is the value of the mass,[5] and a is the value of the acceleration. To solve for F, of course, you put values for m and a into (1.1). If possible, you should also put in their respective units. Be careful not to mix units. That is, make sure you consistently use the same unit system throughout the entire equation. In SI units, (1.1) gives

$$F = (5.0 \text{ kg})(1.6 \text{ cm/s}^2) = 8.0 \text{ kg·cm/s}^2. \tag{1.2}$$

You may need to convert prefixed units (such as cm here) to the root unit (m) in order to get understandable equivalent units in the final answer. In this case (1.2) becomes

$$F = 8.0 \text{ kg·cm/s}^2 = 8.0 \times 10^{-2} \text{ kg·m/s}^2 = 8.0 \times 10^{-2} \text{ N} = 0.080 \text{ N}, \tag{1.3}$$

where the equivalence between the units kg·m/s^2 and N has been used (see Table 1.1) to put the final answer in the usual SI unit for force, the newton N. [In fact, fundamental equations such as (1.1) are the means for determining the equivalency between derived units and their base-unit form.]

1.3.3   ENGLISH AND OTHER UNITS

Any student living in the United States is familiar with English units. They are commonly used in dealing with everyday objects and measurements (a "ten-pound sack," a "30-inch waist"). In addition, health-care practitioners are accustomed to dealing with hybrid units that have been long accepted and traditionally used in the medical field ("systolic blood pressure of 120 mmHg," and "glucose concentration of 100 mg/dL"). Table 1.3 lists some commonly used English units and medical units.

1.4   CONVERSION FACTORS

Because of the large variety of units encountered, an engineer must be able to quickly and accurately convert any value between different units. Many of the conversion factors needed for this class are tabulated in Appendix A.

The only tricky part of doing a units conversion is making sure you don't use the *inverse* of the correct factor by mistake. For example, let's say you want to convert the measurement of the length

---

[5]Note that the symbol m is used here for the value of the mass. But m is also the symbol for the SI unit meter. Be careful not to confuse these two different uses.

Table 1.3: Some Common English and Medical Units		
Quantity	**Unit**	**Symbol**
length	foot	ft
mass	slug	slug
time	second	s
force	pound-force	lbf
pressure	pound-force per square inch	psi
power	horsepower	hp
energy	big calorie or food calorie	Cal
volume	liter	L
cardiac output	liters per minute	L/min
mass concentration	milligrams per liter	mg/L
blood pressure	millimeters of mercury	mmHg
bladder pressure	centimeters of water	cmH_2O

l of the tibia from centimeters to inches. A typical measurement gives

$$l = 42.1 \text{ cm.} \qquad (1.4)$$

You know that the required conversion factor, from Appendix A, relating the length of one inch to its equivalent length in centimeters is:

$$1 \text{ in} = 2.54 \text{ cm.} \qquad (1.5)$$

Dividing both sides of (1.5) by 2.54 cm gives [6]

$$\left(\frac{1 \text{ in}}{2.54 \text{ cm}} \right) = 1. \qquad (1.6)$$

The left-hand side of (1.6)—in the parenthesis—is now equal to unity, so it is a conversion factor you can use to multiply the right-hand side of (1.4) without changing the equality of (1.4):

$$l = 42.1 \text{ cm} \left(\frac{1 \text{ in}}{2.54 \text{ cm}} \right) = 16.6 \text{ in.} \qquad (1.7)$$

The key to knowing that you have multiplied by the correct factor in this case is to notice how the original centimeter units cancel in the numerator and denominator of (1.7), leaving the answer in the desired inch units.

[6]In (1.5), the 1 on the left-hand side is an exact number so it can be considered to contain an infinite number of significant figures. As a consequence the right-hand side of (1.5) sets the significant figure limit of the conversion factor. Now to avoid unduly letting the conversion factor determine the significant figures in the final answer rather than the value being converted, make sure the conversion factor is specified to at least as many significant figures as the number being converted. [Note that in (1.10), the value being converted—not the conversion factor—sets the precision of the final answer, as is proper.]

If the *opposite direction* of conversion is desired, the procedure is similar but now the manipulation of the conversion factor (1.5) must be inverted. Say you wanted to convert

$$s = 6.1 \text{ in} \qquad (1.8)$$

into centimeter units. Divide both sides of the conversion factor (1.5) by 1 in to get

$$1 = \left(\frac{2.54 \text{ cm}}{1 \text{ in}} \right). \qquad (1.9)$$

The right-hand side of (1.9) is a new unity conversion factor you can use to multiply (1.8):

$$s = 6.1 \text{ in} \left(\frac{2.54 \text{ cm}}{1 \text{ in}} \right) = 15 \text{ cm}. \qquad (1.10)$$

Again note that the original inch units cancel, leaving the answer in the desired centimeter units. By keeping careful track of the units in all equations, important verification is provided that the conversion factors—(1.6) and (1.9)—are in the correct order, respectively, to accomplish their intended direction of conversion.

1.4.1   THE USE OF WEIGHT TO DESCRIBE MASS

In the English system, it is common to describe the mass of some object, say a person, by words such as "a one-hundred-and-seventy-pound man." But this description is actually based on *weight* (the same as force—175 lbf in this case) *not* mass, so it is only an indirect measure of the man's mass. In fact, weight is a somewhat untrustworthy way to describe mass since the same mass will have different weights depending upon the local constant of gravitational acceleration. For example, on the moon the man will weigh only 29 lbf.

To convert from weight to mass, one needs to specify the assumed acceleration of gravity. At the earth's sea level, the acceleration of gravity is

$$g = 32.174 \text{ ft/s}^2 = 9.8067 \text{ m/s}^2. \qquad (1.11)$$

(These values are normally the constants assumed for gravitational acceleration if no explicit value of g is given.) Newton's second law, (1.1) with g replacing a, can be used to find the mass of the man in question in English units:

$$m = F/g = (175 \text{ lbf})/(32.174 \text{ ft/s}^2) = 5.44 \text{ lbf·s}^2/\text{ft} = 5.44 \text{ slug}. \qquad (1.12)$$

1.5   CONSISTENCY CHECKS

Mistakes in calculations happen. Of course, careful students and engineers try to keep errors to a minimum, but when they do happen it is very useful to have a set of quick checking tools that alert that something is amiss in the answer. **Consistency checks** are one class of such error-catching tools. These procedures will not necessarily tell you the correct answer, but once you become familiar with their use they will flag mistakes in calculations and equations. Three consistency checks are described below.

1.5.1   REALITY CHECK

Some students call this the "duh" test. It is very quick and easy to perform. You simply think about the value of the answer you just derived, and based on commonsense you make a judgement about whether the answer could possibly be right or not. For example, let's say you calculate the blood pressure in your capillary bed based upon realistic physiological values of capillary compliance and blood volume. After converting to English units, you get the answer

$$P = 475 \text{ psi.} \tag{1.13}$$

If you think about this value for a moment, it is way out-of-line physiologically. The air pressure in your car tires is about 30 psi, so a blood pressure of 475 psi would undoubtedly blow out all your capillaries. You probably made a conversion factor error (incorrectly inverting the conversion factor?).

Of course, the more familiar you are with the units being used, the more errors this method will pick up. SI units, especially newtons and pascals, are still not a part of everyday life. Also, the utility of this method will improve as you gain more experience with values normally encountered for the case being studied.

1.5.2   UNITS CHECK

When you carry units throughout a calculation, the units of the final answer must be consistent with those expected for the dimension of the solved quantity. This is a powerful check. For example, let's say you've solved for the length increase Δl of a 10-cm long bone when a pressure of 18 Pa is applied longitudinally to one end. You come up with an answer of

$$\Delta l = 0.016 \text{ N.} \tag{1.14}$$

Obviously something is wrong, because newton is not an appropriate unit for a length (or a length change); it must be given in units of meter, or millimeter, or inch, or similar. You need to go back and recheck the steps leading to this answer.

Care must be taken when applying this check, though, to make sure you aren't fooled by *equivalent* units. For example, in another calculation of a length, you may get an answer of

$$l = 2.32 \text{ N·s}^2/\text{kg} . \tag{1.15}$$

The units here are actually okay for length, because when the base-unit equivalent to a newton, namely kg·m/s^2 (see Table 1.1), is substituted into the units above, the result is

$$\text{N·s}^2/\text{kg} = \text{kg·m·s}^2/(\text{s}^2 \cdot \text{kg}) = \text{m,} \tag{1.16}$$

which is appropriate for a unit of length.

Units checking is especially helpful in keeping track of which conversion factor to use. You must make sure that the original units cancel, leaving the desired units as discussed in Section 1.4 above. [A units check might have caught the mistake leading to the wrong answer of (1.13).]

1.5.3   RANGING CHECK

This check is most useful for finding errors in derived equations. It involves letting one of the variables on the right-hand side of the equations (an independent variable) take on increasing or decreasing values, even to the limit of infinity or zero, and noting whether the effect on the variable on the left-hand side (the dependent variable) is in the right direction and, in the case of a limit, converges to the correct value. Usually there are several independent variables; any one can be selected for "ranging" in this manner and checked for the expected effect on the dependent variable.

For example, let's say you have solved for the pressure drop ΔP across a blood vessel whose diameter is D and whose length is l by using Poiseuille's Law (to be discussed in Chapter 3) and you get

$$\Delta P = \frac{2\pi D^2 \mu l}{Q}, \qquad \text{(error)} \qquad (1.17)$$

where Q is the volumetric blood flow and μ is the blood viscosity. To quickly check for a major error, do a ranging check on (1.17) by imagining that the independent variable for diameter D takes on larger and larger values. From (1.17), pressure drop ΔP will increase for this case. But it doesn't make sense that a *larger* vessel diameter will produce a *larger* pressure drop (all other parameters staying constant); instead, the pressure drop should get *smaller*. This is a strong indication that a mistake involving D was made in the derivation of the equation.

Doing more with this example, let the value of the vessel length l go to the limit of zero, and by examining (1.17) find the resulting limit of ΔP:

$$\lim_{l \to 0} \Delta P = 0. \qquad (1.18)$$

This result *does* make sense: If the length of a blood vessel gets vanishly short, the pressure needed to force blood flow through it should also vanish. You can conclude, at least, that the location of the variable l in the numerator of (1.17) is correct. [Going further, ranging checks done on the other independent variables shows that the behavior of ΔP upon the ranging of the viscosity variable μ seems reasonable, but the position of the flow variable Q in the equation is questionable. A units check on (1.17) would also have shown that there is an error in the equation.]

1.6   ORGANIZATION OF THE REMAINING CHAPTERS

The remainder of this book is organized in chapters. Each chapter covers an important law or principle relating to biomechanics (fluids and solids), cardiovascular mechanics, electricity, or biosignals. Each chapter has a homework set that will be assigned. Many, but not all, of the chapters relate to the Major Project, and the skills learned in doing the homework will be essential in solving the Major Project.

1.7   PROBLEMS

1.1. As team leader of an engineering group that has developed a new replacement aortic valve, one of your tasks is to calculate the manufacturing cost (parts and labor) of each valve. The cost of each valve has three components:

a. Outside ring: cost = $134.31.

b. Leaflets, made in Italy: 3 needed per valve. Each leaflet costs 88,221 Italian lira. (Exchange rate $1 = 2131.05 Italian lira)

c. Labor: 0.40 hrs at a rate of $132/hr. Keeping the appropriate number of *significant figures* in each step, calculate the cost of each valve.

[ans: $312]

1.2. In a physics class, you have derived the following equation that gives the time t that a body takes, starting from rest, to fall a distance d under the influence of gravity:

$$t = \sqrt{\frac{2d}{g}},$$

where g is the acceleration of gravity (9.8067 m/s^2).

a. Perform a units check on the above equation.

b. Perform one ranging check on the above equation.

CHAPTER 2

Darcy's Law: Pressure-Driven Transport Through Membranes

2.1 INTRODUCTION – BIOLOGICAL AND MAN-MADE MEMBRANES

Cells are the building blocks of all living things, including plants and animals. Cells contain the components and perform the functions that allow life, such as energy conversion, chemical regulation, reproduction and repair. They occur in numerous varieties in different tissues of the human body depending upon the function performed by that tissue. During embryonic development, cells start undifferentiated, and then develop into specific differentiated types appropriate for the various roles they play in the tissues. Thus, the shapes and sizes of cells in the human body vary widely from very small (as thin as 0.2 μm across for the plate-like endothelial cells lining blood capillaries) to moderate (250 μm diameter for a spherical ovum) to very long (nearly 1 m for the threadlike axon projections of some central nervous system cells).

The contents of the cells are packaged inside by the cell wall. A very simplified view of this compartmentalization is shown in Fig. 2.1. In human cells this wall is known as the **plasma membrane** and is composed of a sandwich of two layers of phospholipid molecules facing each other (a "bilayer"). The molecules in the phospholipid bilayer are arranged such that the "water-avoiding" (hydrophobic) ends of the molecules all face inward into the membrane center, while the "water-loving" (hydrophilic) ends all face toward the surface of the membrane. This means that both exposed sides of the plasma membrane are hydrophilic, consistent with the fact that both the fluid surrounding the outside of the cell and the material inside the cell (the **cytoplasm**) are composed mostly of water. Almost all cells in the human body (except for red blood cells) have a nucleus inside containing the DNA for cell replication; these cells are known as **eukaryotic** cells. More primitive cells such as bacteria do not have nuclei and are termed **prokaryotic** cells.

The plasma membrane of the cell serves several purposes. It is foremost a mechanical barrier that compartmentalizes and protects the contents of the cell, separating it from its neighbors. But this barrier cannot be absolute and impenetrable. Otherwise nutrients (e.g., oxygen and glucose) needed to meet the energy requirements of the cell and keep it alive could not enter the cell, and the waste products (CO_2, urea) could not leave. Moreover, specific free ions (sodium, chloride, calcium and potassium to name a few, depending upon the cell function) must be able to travel through the membrane between the cytoplasm and the extracellular space in order to maintain electrical and chemical balance and to carry out the functioning of the cell. Water also passes through the

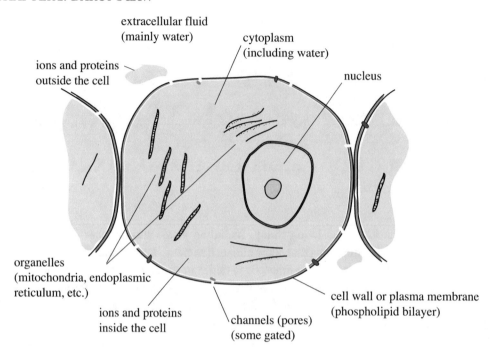

Figure 2.1: A simplified drawing of a generic human cell showing cell contents packaged inside a selectively permeable cell wall—the plasma membrane.

membrane, either from inside to outside or visa versa, depending upon the direction of net force driving the flow.

Therefore, the cell membrane does indeed selectively allow certain molecules (which are specific to the role of the cell) to pass through, thereby regulating cell contents. The cell membrane is said to be **selectively permeable**, or **semipermeable**. The ease with which each species can pass through varies depending upon the size, shape and electrical charge of the species, and upon the characteristics of the membrane. The presence of small channels (pores) through the plasma membrane—some of which are gated—accounts for some of the ability of substances to pass through. Proteins spanning across the membrane can also facilitate transport of selected molecules.

There are three main ways that substances can be transported through a plasma membrane:

1. **By fluid (or hydrostatic) pressure** (the topic of this chapter)–This mechanism is relevant to liquid molecules such as water, and to gases.

2. **By electro-chemical diffusion** (covered in Chapter 14, Part II)–This mechanism applies to ions, small molecules and some macromolecules as well as to water. When water is driven across a semipermeable membrane due to a difference in water concentration on each side

(in turn due to a difference in solute concentrations), the water is said to be driven across by **osmotic pressure**.

3. **By active transport** by membrane-spanning proteins (also covered in Chapter 14, Part II)–This mechanism facilitates the transport of ions, small molecules and macromolecules.

All three mechanisms are in play with cells, but some are more pronounced than others depending upon the cell type. For example, nerve cells rely upon the cooperation between electrochemical diffusion and active transport of ions to maintain their nerve cell function and produce action potentials (Chapter 14, Part II). Water transport by hydrostatic pressure is minimal in these cells.

On the other hand, the cells that line the blood capillaries in the glomerulus of the kidney allow water to pass through their walls relatively easily by the action of hydrostatic pressure from the blood volume into urine-collecting spaces of the kidney. (Some water also leaks out through small gaps in the junctions between the cells.) Osmotic pressure opposes this water movement, but the hydrostatic pressure is greater here, so there is a net flow of water out of the blood in this part of the kidney (most is reabsorbed later in the kidney). The glomerulus thus plays a major role in the regulation of water in the body.

Similarly, the endothelial cells that line the capillaries in the circulatory system also allow some water to pass outward from the blood into the interstitial space outside the vessels, again driven by hydrostatic pressure. The volume of water that leaks out depends upon how much greater the hydrostatic pressure is than the osmotic pressure opposing it. If water secretion is normal, only a small amount of water filters out; the lymphatic system collects it and returns it back into the veins. But if the hydrostatic pressure in the capillaries is abnormally high, caused for example by a weak left heart that doesn't empty the veins readily, the amount of water driven out of the capillaries can overwhelm the lymphatic system, leading to a pooling of water and swelling (edema) of the surrounding soft tissue. When this happens in the capillaries of the legs, water swells the leg's tissues, an early sign of a weak heart. In the lung capillaries, it leads to the collection of water in the lung alveoli (a symptom of "congestive heart failure"), with serious consequences on breathing ability.

2.1.1   MAN-MADE MEMBRANES

In addition to naturally occurring membranes, there are several examples of man-made semipermeable membranes. These can be as simple as the cellulose-fiber paper filters common in chemistry labs. Of more complexity, the first successful artificial kidney, or dialysis machine, was assembled by Dr. Willem Kolff [1] in 1944 using long tubes made with thin cellophane walls. Blood drawn from the veins of a patient was passed through the inside of the tubes, which were immersed in a bath of fresh electrolyte solution. Water, ions and waste products were exchanged across the tubing, clearing

---

[1] Dr. Kolff (1911-2009) had a distinguished career in artificial organs, and joined the faculty of the University of Utah in 1967.

the blood as it was continuously returned to the patient. Figure 2.2(a) is a picture of an early dialysis machine.

a.

b.

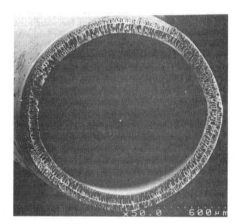

Figure 2.2: Examples of the use of man-made membranes. (a) Dr. W. Kolff shown with an early artificial kidney machine. (b) Cross-sectional view of a small porous tube for encapsulating cells or drugs in host tissue. (Photos courtesy of the Dept. of Bioengineering, University of Utah.)

More recently, tissue bioengineers have fabricated semipermeable membranes for encapsulating collections of cultured cells or drugs. When implanted into body tissue, these membranes form a container keeping the cells together while selectively allowing exchange of nutrients and desirable products from the cells. When the membrane package contains drugs, it allows slow release of the drug in a measured fashion for predicable delivery to the patient. Figure 2.2(b) shows an example of a man-made semiporous membrane.

2.2   DARCY'S LAW

The mathematical relationship between the flow of fluid through a porous obstruction and the pressure driving the flow was first derived by French hydraulic engineer Henry Darcy in 1856. Darcy was engaged in designing a water treatment system for the city of Dijon when he experimented with various flow rates of water through different lengths of tubing filled with sand. The water was driven through the sand by gravity. He found a linear relationship between the flow rates and the driving pressure, and an inverse relationship with the length of the column of sand. Although Darcy's original work was done with water and sand, his findings can be applied to more general porous materials such as membranes.

Darcy's relationships can be illustrated by the arrangement shown in Fig. 2.3. Here a membrane of thickness h (SI units: m) and area A (m^2) is inserted in a fluid-filled tube. A pressure P (Pa) is imposed on the fluid, causing it to flow with flow rate Q (m^3/s) through the membrane.

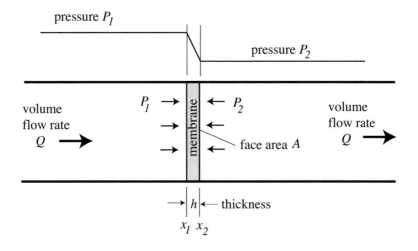

Figure 2.3: Arrangement in which the flow rate Q through the porous membrane is measured as a function of the pressure drop $\Delta P = P_1 - P_2$ across the membrane.

Hydraulic pressure is defined as force per unit area:

$$P = F/A, \tag{2.1}$$

where P (Pa) is the pressure, F (N) is the total force acting on the membrane face, and A (m^2) is the area of the membrane's face. It can be shown easily from (2.1) that the SI unit Pa is equivalent to N/m^2.

The fluid pressure on the left side of the membrane in Fig. 2.3 (the "entrance" side) has a value of P_1. The pressure drops linearly across the membrane (the distribution is shown by the upper solid line in the figure) to a lower pressure P_2 on the right ("exit") side. The pressure drop across the membrane is defined as the difference between the entrance and exit pressures:

$$\Delta P = P_1 - P_2. \tag{2.2}$$

If we now measure the relationship between flow rate Q and the variables in the setup, particularly the pressure drop, we get a plot similar to that shown in Fig. 2.4. The solid line, which

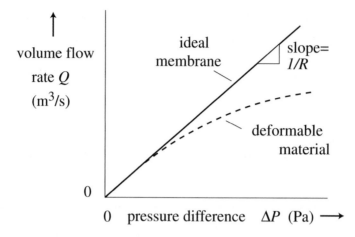

Figure 2.4: Relationship between fluid flow rate and applied pressure difference for two different porous materials. The solid curve shows an ideal porous membrane following Darcy's Law; the dashed curve shows a deformable (nonideal) material.

is representative of the behavior of an ideal membrane, shows a linear relationship between the flow rate and the pressure difference, and can be expressed by Darcy's Law:

$$Q = \frac{kA}{\mu} \frac{\Delta P}{h}, \qquad \text{Darcy's Law} \qquad (2.3)$$

The natural resistance of the fluid to flow is specified by the value of its fluid **viscosity** μ (SI units: kg/m·s). The ease or difficulty with which a particular membrane allows the fluid to pass is given by the **permeability constant** k (m^2) of the membrane. The viscosities of various fluids and the permeability constants of certain materials, including biological materials and filters, are listed in tables in Appendix B.

Note from (2.3) that the membrane properties k, A, and h along with the fluid viscosity μ determine the resistance of the membrane to allow fluid flow. It is reasonable, then, to think of the membrane as opposing the flow of the fluid with a **hydraulic (fluid) resistance** R, which we define as

$$R = \frac{\mu h}{kA}. \qquad (2.4)$$

Inserting (2.4) in (2.3) gives an alternate and more compact form of Darcy's Law:

$$Q = \frac{\Delta P}{R}. \qquad \text{Alternate Form of Darcy's Law} \qquad (2.5)$$

Note from (2.5) that the higher the membrane's resistance is (for example, if the membrane thickness increases), the lower the flow rate is for a given pressure difference.

2.2.1 IDEAL AND NONIDEAL MATERIALS

Darcy's linear relationship applies to many biological and man-made porous materials of interest, especially at low flow rates and pressures. In such ideal materials, the permeability k is a constant and is independent of fluid flow rate, as shown by the solid curve of Fig. 2.4. This applies to many materials, but not all. For example, the dashed curve in Fig. 2.4 shows the relationship between the applied pressure difference and flow rate of a saline solution through articular cartilage in the knee. Although this material obeys a linear law for low flow rates, it is deformable such that as the pressure increases, the material compresses, narrowing or closing off some of the fluid microchannels. This causes its permeability constant k to decrease in the upper portion of the curve, consequently increasing its fluid resistance R to flow. It therefore exhibits a nonideal, nonlinear behavior as opposed to the linear behavior of an ideal material.

Example 2.1. Flow Through a Membrane

A round membrane 2.00 mm thick has a permeability constant of $k = 3.50 \times 10^{-12} m^2$. Find the diameter that would allow water to flow at a volume flow rate of 17.5 cm^3/min with a pressure drop of 0.100 psi.

Solution

First convert the volume flow rate of 17.5 cm^3/min to SI units of m^3/s:

$$Q = 17.5 \frac{cm^3}{min} \left(\frac{1 \times 10^{-6} \ m^3}{1 \ cm^3} \right) \left(\frac{1 \ min}{60 \ s} \right) = 2.92 \times 10^{-7} \frac{m^3}{s}.$$

Also convert the pressure drop to units of Pa:

$$\Delta P = 0.100 \ psi \left(\frac{6895 \ Pa}{1 psi} \right) = 6.90 \times 10^2 \ Pa.$$

From (2.5) the resistance of the membrane needs to be

$$R = \frac{\Delta P}{Q} = \frac{6.90 \times 10^2 Pa}{2.92 \times 10^{-7} m^3/s} = 2.36 \times 10^9 \frac{Pa \cdot s}{m^3}.$$

Putting this value in (2.4) and solving for A, using the value for μ of water from Appendix B, gives

$$A = \frac{\mu h}{Rk} = \frac{(0.0010 \ Pa \cdot s) \left(2.00 \times 10^{-3} m \right)}{\left(2.36 \times 10^9 Pa \cdot s/m^3 \right) \left(3.50 \times 10^{-12} m^2 \right)} = 2.4 \times 10^{-4} m^2.$$

Since $A = \pi D^2/4$, the diameter of the membrane is

$$D = \left(\frac{(4)(2.4 \times 10^{-4} \text{m}^2)}{\pi} \right)^{\frac{1}{2}} = 0.017 \text{ m} = \textbf{1.7 cm}.$$

2.3 MECHANICAL FILTRATION (SIEVING)

In addition to partially resisting the flow of fluid through them, membranes also act as blocking filters, i.e., *cutoff filters*. This is where the filter completely excludes the passage of particles larger than a certain size regardless of the pressure applied. The maximum particle size allowed to pass corresponds roughly to the size of the channels or pores extending through the membrane, specified as the cutoff size or "effective" pore size of the filter. This mechanical filtration capability, also known as sieving, is perhaps the most common use of man-made membranes in the medical and bioengineering lab. Figure 2.5 is a photomicrograph of a typical man-made cellulose-fiber filter showing the intertwined fibers that provide the sieving action. The effective pore size is largely determined by the spacing between the fibers. Other filters can be made from glass fibers or polymer membranes such as nylon or Teflon.

As an example, if the effective pore size of a certain filter is 0.1 μm, no spherical particle with a diameter greater than 0.1 μm can pass through, regardless of the pressure applied—at least until the pressure is so great that it ruptures the membrane. However, for more realistically shaped biological particles that come in a variety of three-dimensional shapes, the situation is more complicated. A long, cylindrically shaped bacterium oriented with its long axis parallel to the channel axis might slip through if its narrowest diameter is below the filter's cutoff size, even though its length may exceed this size. (Nevertheless, for simplicity in calculation, biological particles are sometimes modeled as though they were spherical in shape, specified by a single diameter.) Biological particles are often electrically charged as well, further affecting their ability to pass through the tortuous channels of a filter.

Bioengineers usually characterize the "size" of biological particles or molecules by their mass in terms of **molecular weight** MW. The traditional (non-SI) units of molecular weight are **Daltons** (abbreviated Da) in honor of English chemist John Dalton. One proton or one neutron have essentially the same weight; that sets the size of one Dalton [2]:

$$\boxed{1\text{Da} = 1.661 \times 10^{-24} \text{ g}} \tag{2.6}$$

Thus, a neutron or proton has a mass of ~ 1 Da while an electron is much lighter (9.1×10^{-28} g) and thus has a mass only about 1/2000 Da. Proteins are much larger. For example, the blood protein serum albumin has a molecular weight of 68 kDa, and the blood protein Immunoglobulin M has a molecular weight of 1000 kDa.

---

[2]The Dalton was initially defined as 1/16 the weight of an oxygen atom (nominally 8 protons, 8 neutrons and 8 electrons) in a mixture of isotopes of oxygen. The more recent atomic mass unit (amu or u) is only very slightly different in weight from a Dalton.

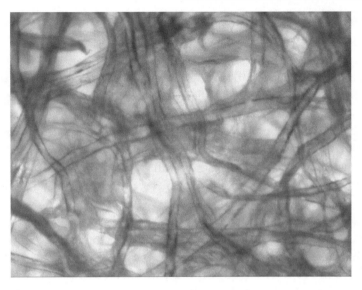

Figure 2.5: View under a microscope of the intertwined cellulose fibers in man-made filter paper. The pore size is roughly determined by the opening between fibers; for this filter, the effective pore size is 11 μm.

Proteins—biological particles commonly found in body fluids—come in a variety of sizes and shapes. Still, since the subunits that make up proteins take up about the same space independent of the particular protein they belong to, the mass density of all proteins is about the same, 1.37 g/cm^3. In a similar fashion, the mass density of DNA is reasonably uniform among different DNA strands, about 1.6-1.7 g/cm^3. Viruses, which are composed mostly of a DNA core enclosed in a thin protein coating, have about the same density as DNA. In summary, to a good approximation, the density of each of these particles is:

$$\rho_{\text{protein}} \approx 1.37 \text{ g/cm}^3$$
$$\rho_{\text{DNA, virus}} \approx 1.6 - 1.7 \text{ g/cm}^3. \tag{2.7}$$

Once the density and molecular weight of the particle are known (and assuming a spherical shape for simplicity), the radius r of the molecule can be estimated by the following steps. The total mass M of the particle is given by its molecular weight in Dalton units times the equivalent weight of one Dalton:

$$M = \text{MW} \cdot (1.661 \times 10^{-24} \text{ g / Da}). \tag{2.8}$$

The density ρ is total mass divided by volume V:

$$\rho = M / V, \tag{2.9}$$

and the volume of a sphere is

$$V = \frac{4}{3}\pi r^3. \tag{2.10}$$

Using (2.7), (2.8), (2.9), and (2.10) together, the radius of a particular particle can be estimated (see the example below, which estimates the diameter of serum albumin, a protein, to be 5.4 nm). Viruses can be as small as 10 nm in diameter, but are generally on the order of 100 nm in diameter; some forms of the human influenza virus are about 80-120 nm in diameter. Bacteria (which are prokaryotic cells) are usually much larger than viruses, with a diameter range of about 1-10 μm. Eukaryotic (plant and animal) cells are larger still, with typical diameters in the 10-100 μm range. A red blood cell is a biconcave disk with a disk diameter of about 8 μm and a thickness of about 2 μm.

As mentioned above, porous membranes are often used to filter out particles larger than a certain size. Thus, a backpacking water filter can be used to eliminate bacteria and spores from drinking water while still allowing the small water molecules (which are about 0.3 nm in diameter) to pass. A cutoff pore size of 0.4 μm (400 nm) will filter out bacteria such as E. coli, typhoid, and cholera. Similarly, the eggs and larvae of parasites are about 20-100 μm in size, giardia cysts are about 14 μm, and cryptosporidium is about 4 μm in size, so these pathogens would be filtered out as well. Viruses and proteins, on the other hand, are so small it is difficult to make a practical water filter for them; the pore size would be so small that the filter's permeability to water molecules would be very low, making water passage difficult at reasonable pressures. For viruses, alternative treatment methods such as iodine or boiling the water are often used.

Example 2.2.   Diameter of a Protein
Find the diameter of the protein serum albumin, given that its molecular weight is 68 kDa.

Solution
Since the density of all proteins is approximately 1.37 g/cm^3, (2.8) and (2.9) together give the protein volume as

$$V = \left(\frac{6.8 \times 10^4 \text{ Da}}{1.37 \text{ g/cm}^3}\right)\left(\frac{1.661 \times 10^{-24} \text{g}}{1 \text{ Da}}\right) = 8.2 \times 10^{-20} \text{ cm}^3.$$

Then using (2.10) the radius r is found to be

$$r = \left(\frac{(3)(8.2 \times 10^{-20} \text{ cm}^3)}{4\pi}\right)^{1/3} = 2.7 \times 10^{-7} \text{ cm} = 2.7 \text{ nm}.$$

Thus, the diameter [3] is about **5.4 nm**.

---

[3] Remember, diameter is twice the radius. Also, this calculation assumes a spherical shape for the protein. Actually, serum albumin is more cylindrical in shape.

2.4   PROBLEMS

2.1. a. Using Equation (2.4), find the SI units for fluid resistance.

$$[\text{ans: } kg/(s \cdot m^4) \text{ or } Pa \cdot s/m^3]$$

b. Using Newton's second law, Equation (1.1), show that $kg/(s \cdot m^4)$ is equivalent to $Pa \cdot s/m^3$.

c. Using the above results, do a units check on Equation (2.5) to show that the units on the left-hand side are consistent with the units on the right-hand side.

d. A common non-SI unit of viscosity is the poise (P), which is equivalent to $g/(cm \cdot s)$. Using conversion factors and the results of part b, show that $1 \text{ P} = 1 \times 10^{-1} \text{ Pa} \cdot s$.

2.2. A certain blood plasma filter has a permeability constant of $1.0 \times 10^{-13} \text{ m}^2$. Its area is 1.0 cm^2 and its thickness is 1.0 mm. When a net pressure of 10.00 psi is applied across the filter, what is the volume flow rate of blood plasma through the filter? If spherical drops of blood plasma, each of diameter 4.0 mm, come out of the filter, how many drops per second will flow?

$$[\text{ans: } Q = 5.7 \times 10^{-7} \text{ m}^3/s; \text{ 17 drops/s}]$$

2.3. Estimate the molecular weight of a blood protein that is approximately spherical in shape with a diameter of 20.0 nm.

$$[\text{ans: } MW \approx 3460 \text{ kDa or } 3.46 \times 10^6 \text{ Da}]$$

2.4. A certain backpacking water filter has an "effective" pore size (i.e., a pore diameter) of $2.0 \ \mu m$.

a. Will this filter trap the hepatitis A virus, which has a total molecular weight (including protein envelope) of about 32,000 kDa? (Calculate the virus diameter assuming it is spherical and using a reasonable estimate of its density.) If it is not trapped, how could you kill the virus in the water when backpacking?

$$[\text{ans: } Diam \approx 40 \text{ nm, so it is not trapped.}]$$

b. Will this filter trap the giardia cyst? If not, how could you kill giardia in the water when backpacking? (Note: you may find it useful to search the internet or reference books for the answers to parts b and c of this problem.)

c. Is the *E. coli* O157:H7 bacterium a "good" bacterium, or a "bad" bacterium? What effects does it have on the body? Will this filter trap the *E. coli* O157:H7 bacteria? If not, how could you kill these bacteria in the water when backpacking?

2.5. **a.** Estimate the total number of hepatitis A viruses that could fit (in one layer) inside the period at the end of this sentence. Use the virus diameter found in Prob. 2.4a.

[ans: Approx. 130,000,000. Your answer may vary depending
upon your measurement of the size of the period.]

b. Repeat this estimate for red blood cells. Assume they all lie flat in the plane of the paper.

[ans: Approx. 3100.]

CHAPTER 3

Poiseuille's Law: Pressure-Driven Flow Through Tubes

3.1 INTRODUCTION – BIOLOGICAL TRANSPORT

Diffusion is a major mechanism for transporting vital molecules into and away from cells (as mentioned in Chapter 2 and to be covered in more detail in Chapter 14, Part II). For single cell organisms, such as bacteria, or for multi-cell organisms with only a few cells, the distances from inside the cell to the outside environment are short enough that diffusion times for these organisms are sufficiently rapid to sustain life. For example, an oxygen molecule will travel across 100 μm (the size of a moderately large cell) in a water environment in only about 1 second.

But for larger organisms, certainly for animals including humans, diffusion times are much too long to rely on diffusion alone to transport life-supporting molecules throughout the entire body. Einstein derived the following relationship describing the average time needed for a molecule to diffuse a distance d driven by a planar concentration gradient:

$$t = \frac{d^2}{4D}, \tag{3.1}$$

where t is the average time (s) for a molecule to diffuse a distance d (m), and D (m^2/s) is a diffusion constant that depends upon the molecular size, shape and charge, the viscosity of the surrounding medium and the temperature. Note that the diffusion time varies as the *square* of the distance traveled. Thus, for an oxygen molecule to travel from the human lung (where indeed it is rapidly absorbed into the blood across the thin alveolar walls by diffusion) to a far-reaching part of the body, say the foot, would take about 6 years to travel by diffusion alone! Therefore, some means of augmenting diffusion by a bulk fluid flow—i.e., blood flow—is necessary. This is a major role of the circulatory system in larger organisms: to expeditiously carry oxygen and other nutrients from the outside world to the remote tissues of the organism, and then to transport waste products back out. (Blood flow also facilitates heat exchange and immunological defenses in the body.)

But oxygen by itself is not readily dissolved in pure blood plasma (mostly water), so some molecular carrier is needed. This task falls to hemoglobin, which effectively binds oxygen in the lungs (becoming oxyhemoglobin), then releases it where needed in tissues at lower oxygen pressure. Hemoglobin molecules are packaged inside small biconcave disk-shaped red blood cells (erythrocytes) about 8 μm in diameter; together the red blood cells and the blood plasma comprise **whole**

blood. The percentage of whole blood volume taken up by the red blood cells gives the **hematocrit** of the blood—normally between 40-45% for humans, a little higher in males than in females.

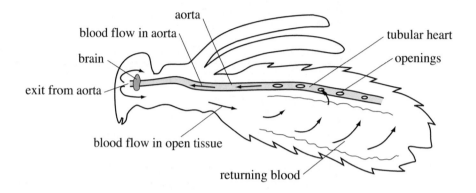

Figure 3.1: Open circulation in insects. After it leaves the heart, the blood empties into the entire body volume. It is then collected through openings in the heart to be re-pumped. (After Withers, 1992.)

To pump the blood throughout the body of the organism, some pressure source is needed (a heart), and some arrangement for directing outward flow and collecting the spent blood is required (a circulatory system). Different levels of insects and animals have developed various strategies for their heart and circulation. Figure 3.1 shows the circulatory arrangement common in insects. This is an **open** circulation, meaning that the blood pumped by the single tubular heart (which periodically expands and contracts its diameter by muscles attached to its circumference) is released into the open field of tissue filling the insect's body rather than being contained in arteries, capillaries or veins. It perfuses the open tissue and picks up new oxygen before reentering the heart through openings in its walls to be re-pumped. This system works well for small organisms, but in larger animals the open field of tissue would present an uneven and inefficient distribution for flow.

In fish (see Fig. 3.2) the system is **closed** such that the blood is contained in a continuous network of vessels passing through the body. The fish heart is two-chambered (an atrium preceding the stronger ventricle) but the heart is **one-sided**. This means that blood expelled out of the fish heart passes first through the gills to pick up oxygen, then directly to the rest of the tissues without any intervening pressure boost, before returning back to the heart via the veins.

In mammals, including man, the circulatory system is similarly closed but the circulation is divided into two segments connected in series. Thus, the heart is **two-sided**, each side being the pump for its respective segment of the system. As shown in Fig. 3.3, the right side of the heart (comprised of two chambers, the right atrium and right ventricle) sends blood through the lung capillaries at relatively low pressure where the red blood cells pick up oxygen; this part of the system is called the **pulmonary** circulation. The oxygenated blood then returns back to the left side of the heart (also comprised of two chambers) where it is pumped with much higher pressure (about five

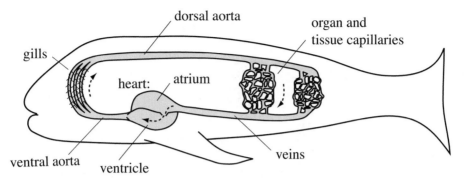

Figure 3.2: Closed circulation of fish, in which the blood remains inside tubes. The fish heart has only one side.

times higher than in the pulmonary circuit) through the remaining tissues of the body; this portion is the **systemic** circulation. Since the circulatory system contains both a heart and vessels, it is called the **cardiovascular** (CV) system.

The network of tubing making up the human circulation is complex. After passing out of the outflow valves of the respective ventricles, the blood is initially directed through a large vessel (the aorta in the systemic circulation) before splitting progressively into smaller and more numerous arterial vessels. The arteries in turn split into arterioles (called the "gatekeepers" due to their smooth-muscle walls that can contract or expand in diameter), then into a vast number of thin-walled capillaries (as small as 6-7 μm in diameter) where molecular exchange takes place by diffusion. Although it is not evident from the drawings, capillaries are ubiquitous throughout the body, passing to within about 100 μm (the width of a human hair) of every living cell in the body. After the capillaries, the vascular tree begins to recollect the blood into progressively larger and fewer vessels, first the venules, then the veins, and finally into one or two large return ducts (the vena cava in the systemic circulation) that empty into the atrium on the other side of the heart.

All of these vessels have some elasticity or compliance, especially in the venous portion of the systemic circulation where a good percentage (approximately two-thirds) of the body's total blood volume resides. This elasticity is especially beneficial in smoothing out the pulsing nature of the blood flow from the beating heart, as will be seen in the next chapter, as well as in helping propel the blood along. The volume of blood pumped around the system per minute is known as the **cardiac output**, or CO, with traditional clinical units of L/min.

Thus, it can be seen that the vascular network is composed of tubes of various lengths, diameters and connections, and that blood is driven through the network by the pressure produced by the respective ventricles of the heart. To quantify the amount of flow for a given pressure, we need to study the flow of fluids through tubes, next.

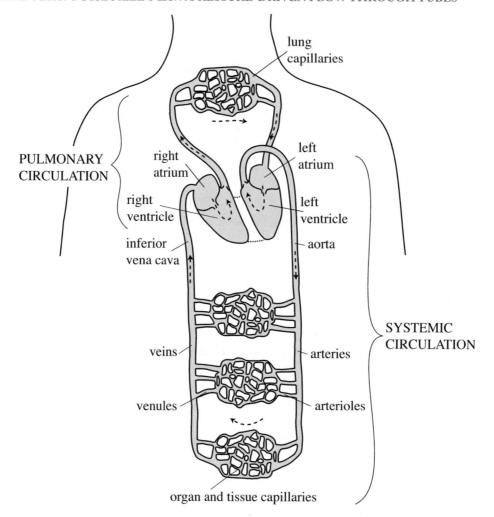

Figure 3.3: A schematic diagram of the human circulation, a closed system. There are *two* sides of the heart: the right side delivers blood to the lungs for oxygenation (the pulmonary circulation); the left side receives this blood and pumps it at higher pressure through the remainder of the body (the systemic circulation). (After Guyton and Hall, 2000.)

3.2   POISEUILLE'S LAW

Between 1838 and 1840, G. Hagen and J. L. Poiseuille independently obtained the relationship between fluid flow in a tube and the pressure required to produce this flow. This relationship is called the Hagen-Poiseuille law, or simply **Poiseuille's Law**. Figure 3.4 shows the arrangement analyzed.

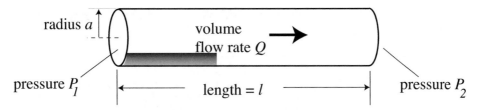

Figure 3.4: Tube for illustrating Poiseuille's Law.

In Fig. 3.4, a volume **flow rate** Q of fluid passes through a tube of **length** l under a **pressure difference** $\Delta P = P_1 - P_2$. Hagen and Poiseuille found the following relationship:

$$Q = \frac{\pi a^4}{8\mu} \frac{\Delta P}{l}, \qquad \text{Poiseuille's Law} \qquad (3.2)$$

where Q = volume flow rate (m^3/s)

$\quad \Delta P = P_1 - P_2 = $ pressure difference (Pa or N/m^2 or kg/m·s^2)

$\quad P_1 = $ fluid pressure at entrance to tube (Pa)

$\quad P_2 = $ fluid pressure at exit from tube (Pa)

$\quad a = $ tube radius (m)

$\quad \mu = $ fluid viscosity (kg/m·s or Pa·s), and

$\quad l = $ length over which the pressure drop is measured (m).

The existence of the various terms in (3.2) can be qualitatively explained by a ranging check using each of the variables in turn (except for the factor 8, which is only found by a mathematical derivation beyond the scope of this chapter). That ΔP appears in the numerator seems reasonable, since for a given length of tube, the higher the driving pressure, the higher the flow should be. The location of l in the denominator also makes sense, since the longer the tube, the less the flow for a given pressure drop (as anyone who has used a very, very long garden hose knows).

The terms πa^4 and μ in (3.2) need a little more explanation. One portion of the πa^4 term (namely πa^2) can be seen simply from the fact that the tube's cross-sectional area is given by πa^2 and volume flow is proportional to cross-sectional area for a given pressure drop. The remaining a^2 dependence requires a look at the fluid flow velocity inside the tube. For Poiseuille's Law to hold, the flow profile is assumed to be **laminar** (or "layered"), as diagrammed in Fig. 3.5. A laminar profile is characterized by a "no-slip" condition at the walls; that is, the fluid velocity is zero where the fluid touches the walls, and then the velocity increases parabolically toward the peak velocity at the centerline of the tube. Note that the velocity must build in a parabolic manner from zero (at the walls) to a peak value (at the center). So the smaller the radius of the tube, the less distance the velocity has to build to a peak value, thus reducing this peak value and the volumetric flow rate. When the two-dimensional nature of the cross-section is considered, the additional a^2 term is found, leading to the overall πa^4 dependence.

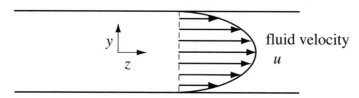

Figure 3.5: Velocity profile for laminar flow in a tube.

This fourth power dependence on radius a is dramatic. It means that if the radius of a tube is reduced by 1/2, the flow rate will be reduced to 1/16th of its original flow! This would severely limit the amount of blood flowing through the tiny capillaries were it not for the fact that there are many, many capillaries in parallel in the vascular network.

The presence of the viscosity term μ in the denominator of (3.2) also needs some discussion. Viscosity is a measure of the resistance of the layers of the fluid to flowing past one another—"sliding friction" as it were. We first saw the term in Chapter 2 relating to flow through membranes. The laminar flow profile in Fig. 3.5 requires that neighboring layers of fluid must slide past each other going from the walls to the centerline, imposing a shearing nature to the flow, as shown for a local region of the fluid in Fig. 3.6.

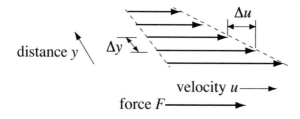

Figure 3.6: Layers of fluid sliding past each other, giving a shear rate $\Delta u / \Delta y$, which requires a force F.

The extent of the shear is expressed in terms of the **shear rate**:

$$\text{shear rate} = \Delta u / \Delta y. \tag{3.3}$$

It takes a force F to overcome the friction of the layers sliding past one another. The force is described locally in terms of the **shear stress** τ:

$$\tau = F/A, \tag{3.4}$$

where A is the area over which the force F acts.

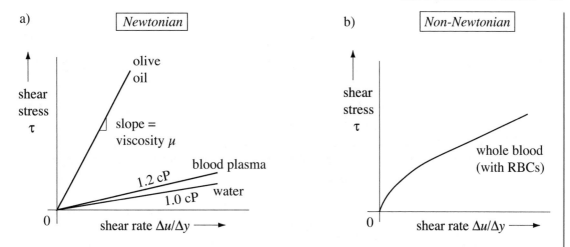

Figure 3.7: (a) Plot of the shear stress vs. shear rate for three fluids that exhibit a linear relationship (Newtonian). The slope gives the fluid's viscosity μ, which is a constant. (b) Plot of whole blood. The red blood cells cause the fluid to be Non-Newtonian, where the viscosity varies with the flow rate.

When the shear stress of a fluid is plotted as a function of the shear rate, a graph similar to Fig. 3.7(a) is obtained for many well-behaved fluids. It shows a linear relationship between shear stress and shear rate, such that the force required to drive the laminar flow is proportional to the velocity of the flow. When put in terms of shear stress τ and shear rate, the fluid obeys the following equation:

$$\tau = \mu \cdot (\Delta u / \Delta y), \tag{3.5}$$

where the proportionality constant μ is the viscosity of the fluid.

Now the presence of the viscosity term μ in the denominator of (3.2) seems reasonable: the more the fluid resists shear flow (i.e., the higher its viscosity), the more pressure ΔP it takes to cause a certain flow rate Q. Molasses takes more force to flow through a tube than water. When the viscosity value is a constant that is not dependent upon the flow rate, as in Fig. 3.7(a), the fluid is termed **Newtonian**.

However, some fluids exhibit a viscosity that varies with flow rate. Such fluids are termed **Non-Newtonian**. An example is whole blood, where the addition of the red blood cells to blood plasma causes the viscosity to increase several fold from that of plasma (so "blood is thicker than water") and also to become variable, as shown in Fig. 3.7(b).

3.2.1   SIMPLIFIED VERSION OF POISEUILLE'S LAW

As with Darcy's Law, we can put Poiseuille's Law (3.2) in a form that involves the resistance of the tube. For flow through tubes, the **fluid resistance** (or hydraulic resistance) is given by

$$R = \frac{8\mu l}{\pi a^4}. \tag{3.6}$$

With this definition of fluid resistance, Poiseuille's Law (3.2) has the following simplified form:

$$\boxed{Q = \frac{\Delta P}{R} = \frac{P_1 - P_2}{R}.}\quad \text{Alternate Form of Poiseuille's Law} \tag{3.7}$$

It is often convenient in diagrams to show flow through pipes or tubes by replacing the physical shape of the tube with a *symbol* representing the resistance R of the tube. The symbol for a fluid element that has resistance to flow, such as a tube, is given in Fig. 3.8.

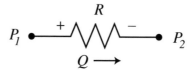

Figure 3.8: Symbol for a resistive fluid element, such as a tube.

3.2.2   ASSUMPTIONS FOR POISEUILLE'S LAW

For Poiseuille's Law to hold, there are a number of assumptions that must apply to the flow through the tube:

- The length of the tube must be much greater than the radius,

- The flow must be steady in time and laminar in velocity profile,

- The fluid must be Newtonian, and

- The tube must be rigid.

Actually, in biological circulatory systems, none of these assumptions are strictly met for flow through the entire organism. However, they hold to some degree under certain conditions. Let's examine the areas of validity for each assumption for blood flow in the human circulation.

- *Length is much greater than radius* – This approximation is valid in the aorta, in longer arteries and in some of the small-diameter, long capillaries, but it is not very valid in the rapidly branching, shorter networks found throughout the system.

- *Flow is steady and laminar* – The blood flow in the aorta and arteries is pulsatile, so the approximation is not valid in those vessels. But when the flow reaches the capillaries and veins, it becomes increasingly steady due to the damping action of the compliant vessels (to be covered in Chapter 4).

Whether the flow profile is laminar or not can be predicted by the flow's **Reynolds number** $\approx R_e$, which is a dimensionless (that is, unitless) number calculated by

$$R_e = \frac{\rho u D}{\mu}, \tag{3.8}$$

where ρ is the fluid density (kg/m^3), u is the fluid velocity (m/s), D is the diameter of the tube (m), and μ is the dynamic fluid viscosity (kg/m·s or Pa·s). Reynolds numbers of 2000-4000 are often considered to be the approximate dividing values between laminar and turbulent flow: when the Reynolds number is below 2000, the flow is likely laminar, while if the Reynolds number is above 4000, the flow will usually be turbulent. Between 2000 and 4000, the flow can be either depending upon other factors such as the time course of the flow and nearby boundary disruptions—e.g., branching—since that will alter the flow profile. Now almost all flows in the human body (with the exception of blood flow in the large aorta during peak ejection from the heart) are of such low flow velocities that the Reynolds number is much below 2000 and the flow is laminar, not turbulent.

- *Fluid is Newtonian*—Without red blood cells, the blood plasma is reasonably Newtonian [see Fig. 3.7(a)], but when the red blood cells are added, the whole blood becomes Non-Newtonian, except over narrow ranges of flow rates [see Fig. 3.7(b)].

- *Tube is rigid*—All blood vessels are not rigid, but rather are distensible and flexible to some degree, with veins being more flexible than arteries. This proves beneficial to the circulation by smoothing out the flow.

So it is clear that the assumptions inherent in Poiseuille's Law do not apply at all times and in all places in the blood circulation. Nevertheless, Poiseuille's Law in the form of (3.2) is still of value when applied to many of the individual vessels. Furthermore, when an entire ensemble of vessels is analyzed (as in calculating the peripheral resistance of the systemic circulation—see example below), detail about the vast number of individual vessels is not known so the simplified form (3.7) is used, in which case a single combined resistance characterizes the ensemble. On this scale, the validity of the assumptions is less important and Poiseuille's Law in the form of (3.7) becomes even more applicable.

Example 3.1. Peripheral Resistance of the Human Systemic System
In physiology terms, the fluid resistance of the entire human systemic circulation, from the inlet at the aorta to the outlet of the vena cava, is called the "peripheral resistance." Given that a typical human cardiac output is 5.5 L/min, that the average pressure in the aorta is 100 mmHg, and that the average pressure in the right atrium is 7 mmHg, find the peripheral resistance in units of mmHg·s/L.

Solution

The peripheral resistance can be modeled with the simplified form of Poiseuille's Law:

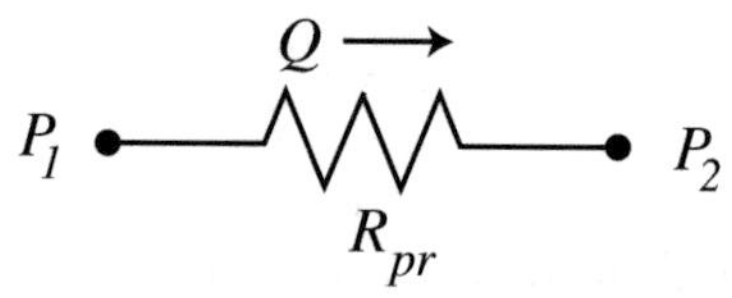

Figure 3.9: Model of peripheral resistance used in Example 3.1

where $Q = \dfrac{\Delta P}{R_{pr}}$, so $R_{pr} = \dfrac{\Delta P}{Q}$. First find Q in units of L/s:

$$Q = 5.5\frac{L}{min}\left(\frac{1\ min}{60\ s}\right) = 9.2 \times 10^{-2}\frac{L}{s}.$$

Also $\Delta P = P_1 - P_2 = 100\ mmHg - 7\ mmHg = 93\ mmHg$. Then

$$R_{pr} = \frac{93\ mmHg}{9.2 \times 10^{-2}L/s} = 1.0 \times 10^3 \frac{mmHg \cdot s}{L} = \mathbf{1000\ mmHg \cdot s/L}.$$

3.3 POWER EXPENDED IN THE FLOW

It takes energy to force the fluid through the tube against its resistance, and since power is defined as the energy expended per unit time, there is a power requirement for maintaining the flow. This power Φ is given by the product of the pressure drop and the volume flow rate:

$$\Phi = \Delta P \cdot Q. \tag{3.9}$$

Relating pressure drop and flow to resistance by (3.7) gives alternate forms for the expended power:

$$\Phi = Q^2 R = \frac{(\Delta P)^2}{R}. \tag{3.10}$$

These power concepts are addressed again in Chapter 7.

3.4 SERIES AND PARALLEL COMBINATIONS OF RESISTIVE ELEMENTS

The circulatory network of vessels contains a great number of interconnections: some vessels can be considered connected in **series** (where the flow goes through each tube sequentially without splitting), others in **parallel** (where the flow splits into several branches before recombining). Using the concept of fluid resistances, a combination of several tubes either in series or in parallel can be easily handled by defining an **equivalent resistance** R_{eq} that represents the effect of the resistances lumped together. These two situations are considered next.

3.4.1   SERIES

This configuration can be illustrated using three hydraulic resistors in series:

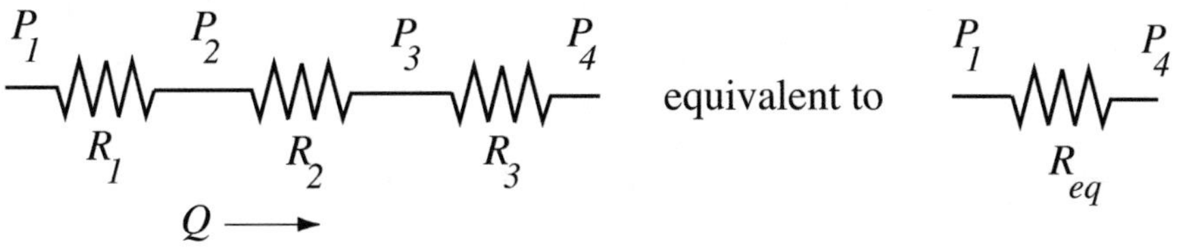

Figure 3.10: Resistive elements in series can be replaced by an equivalent resistor.

Since the same flow Q must go equally through all elements, (3.7) can be applied to each resistance in turn:

$$P_1 - P_2 = R_1\,Q$$
$$P_2 - P_3 = R_2\,Q$$
$$\text{and}\qquad P_3 - P_4 = R_3\,Q. \tag{3.11}$$

Adding these three equations together and defining an equivalent resistance R_{eq} gives

$$\Delta P = P_1 - P_4 = (R_1 + R_2 + R_3)Q = R_{eq}Q, \tag{3.12}$$

where the equivalent series resistance can be seen to be

$$\boxed{R_{eq} = R_1 + R_2 + R_3} \qquad \text{Series} \tag{3.13}$$

and the total pressure drop is related to the flow rate by (3.12). The general formula for combining N resistors in series is

$$R_{eq} = \sum_{n=1}^{N} R_n. \tag{3.14}$$

where Σ is the standard notation for summation.

3.4.2   PARALLEL

The parallel case can be illustrated with two resistive elements in a branching configuration in Fig. 3.11. Here the flow is split—usually unequally—between the two branches. Flow Q_1 goes through resistor R_1 and Q_2 goes through R_2. By the **conservation of volume** principle for incompressible fluids, the total flow Q entering (and exiting) the entire circuit must be equal to the sum of the flows in the two branches:

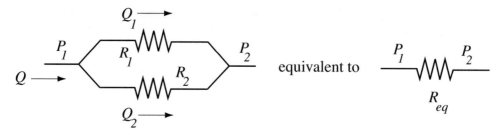

Figure 3.11: Resistive elements in parallel can be replaced by an equivalent resistor.

$$Q = Q_1 + Q_2. \tag{3.15}$$

The pressure drop is the same across each element, given by

$$P_1 - P_2 = \Delta P. \tag{3.16}$$

Using (3.7), the branch flows are related to the common pressure drop by

$$Q_1 = \frac{\Delta P}{R_1}, \tag{3.17}$$

$$\text{and} \qquad Q_2 = \frac{\Delta P}{R_2}. \tag{3.18}$$

Putting (3.17) and (3.18) into (3.15) and defining an equivalent resistance gives

$$Q = \frac{\Delta P}{R_1} + \frac{\Delta P}{R_2} = \Delta P\left(\frac{1}{R_1} + \frac{1}{R_2}\right) = \frac{\Delta P}{R_{\text{eq}}}, \tag{3.19}$$

where the equivalent parallel resistance can be seen to be

$$\boxed{\frac{1}{R_{\text{eq}}} = \frac{1}{R_1} + \frac{1}{R_2}.} \qquad \text{Parallel} \tag{3.20}$$

The general formula for combining N resistors in parallel is

$$\frac{1}{R_{\text{eq}}} = \sum_{n=1}^{N} \frac{1}{R_n}. \tag{3.21}$$

A more complex flow circuit containing both parallel and series combinations can be analyzed by combining each segment using the appropriate equivalent resistances, in order, until the entire circuit is represented by an equivalent resistance. Application of this technique is given in the second example below.

Example 3.2. Adding Tubing to a Membrane Filter

Let the semipermeable membrane analyzed in the example of Chapter 2 (see p. 21) have inlet and outlet tubing added to each end of the membrane, as shown in Fig. 3.12.

Assuming that the flow rate of water through the entire assembly is the same as in the example in Chapter 2 ($Q = 2.92 \times 10^{-7} \mathrm{m}^3/\mathrm{s}$), and assuming that the inside diameter of the tubing is the same as that of the membrane ($D = 1.7$ cm), calculate how much *additional* pressure drop is incurred when the tubing is added.

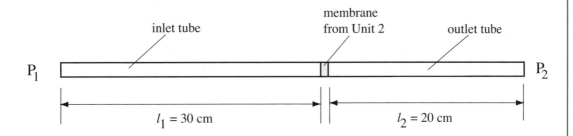

Figure 3.12: Drawing of configuration analyzed in Example 3.2.

Solution

The tubing can be considered to add additional fluid resistance to the overall assembly. The resistance of each tube can be found from (3.6):

$$R_1 = \frac{8\mu l_1}{\pi a_1^4} \quad \text{and} \quad R_2 = \frac{8\mu l_2}{\pi a_2^4},$$

in which $a_1 = a_2 = D/2 = 8.5 \times 10^{-3} \mathrm{m}$.

Since the resistances of the two tubes are in series, their resistances add, as given by (3.14):

$$R_{\text{tubing}} = R_1 + R_2 = \frac{8\mu (l_1 + l_2)}{\pi (D/2)^4} = \frac{8\left(1.0\times10^{-3}\mathrm{Pa} \cdot \mathrm{s}\right)\left(5.0\times10^{-1}\mathrm{m}\right)}{\pi \left(8.5 \times 10^{-3}\mathrm{m}\right)^4} = 2.4 \times 10^5 \frac{\mathrm{Pa} \cdot \mathrm{s}}{\mathrm{m}^3}$$

where the viscosity of water $\mu = 1.0 \times 10^{-3}\mathrm{Pa \cdot s}$ has been used. The additional pressure drop that is added to the original pressure drop (0.100 psi) can be found from (3.7):

$$\Delta P_{add} = Q \cdot R_{tubing} \quad = \quad (2.92 \times 10^{-7} \text{ m}^3/\text{s}) \cdot (2.4 \times 10^5 \text{ Pa} \cdot \text{s/m}^3)$$
$$= \quad 7.0 \times 10^{-2} \text{ Pa} = \textbf{0.000010 psi.}$$

Note that the additional pressure drop due to the tubing is only $1/10{,}000^{\text{th}}$ of the original pressure drop, so it is negligible compared to the drop across the membrane itself. This is because the tubing is relatively short in length and large in diameter.

Example 3.3.   Branching Vessels
Whole blood flows through a single vessel before splitting into two identical parallel vessels in Fig. 3.13.
The length of each segment is 2.0 cm. The combined cross-sectional area of the two parallel vessels is the same as that of the single vessel. Given a blood flow rate of 1.0 mL/min, what is the pressure drop across the entire configuration? (Assume that the viscosity of whole blood is 4.0 cP.)

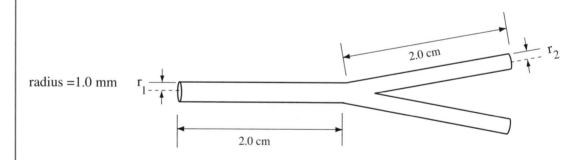

Figure 3.13: Branching vessel arrangement analyzed in Example 3.3.

Solution
This arrangement can be analyzed as one resistance in series with a parallel combination of two other resistances in Fig. 3.14.
We can use Poiseuille's Law to find each of the resistances, after we first find the radius r_2 of each of the identical parallel tubes. Given that the combined area of the two tubes is equal to that of the single tube, then

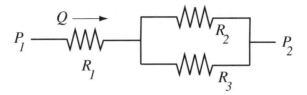

Figure 3.14: Schematic model of the branching tubes.

$$\pi r_1^2 = \pi r_2^2 + \pi r_3^2 = 2\pi r_2^2,$$

so $$r_2 = \frac{r_1}{\sqrt{2}} = \frac{1.0 \text{ mm}}{\sqrt{2}} = 0.71 \text{ mm} = 7.1 \times 10^{-4} \text{ m}.$$

Then, using (3.6),

$$R_2 = R_3 = \frac{8\mu l}{\pi (r_2)^4} = \frac{8\left(4.0 \times 10^{-3} \text{ Pa} \cdot \text{s}\right)\left(2.0 \times 10^{-2} \text{ m}\right)}{\pi \left(7.1 \times 10^{-4} \text{ m}\right)^4} = 8.0 \times 10^8 \frac{\text{Pa} \cdot \text{s}}{\text{m}^3}.$$

Now combine these two resistances in parallel using (3.20). See Fig. 3.15 where $\dfrac{1}{R_p} = \dfrac{1}{R_2} + \dfrac{1}{R_3}$.

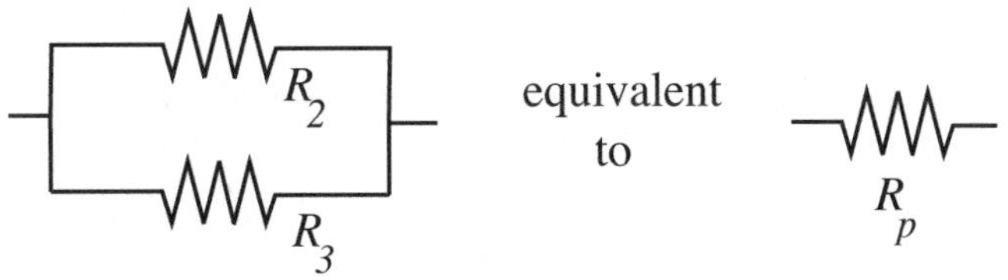

equivalent
to

Figure 3.15: Two parallel tubes combined into one equivalent resistance.

Since $R_2 = R_3$, $\dfrac{1}{R_p} = \dfrac{2}{R_2}$, or $R_p = \dfrac{R_2}{2} = \dfrac{8.0 \times 10^8}{2} = 4.0 \times 10^8 \dfrac{\text{Pa} \cdot \text{s}}{\text{m}^3}$.

Hence, the entire configuration can be modeled as two resistors in series in Fig. 3.16 where $R_{\text{eq}} = R_1 + R_p$. Again using (3.6),

$$R_1 = \frac{8\mu l}{\pi (r_1)^4} = \frac{8\left(4.0 \times 10^{-3} \text{ Pa} \cdot \text{s}\right)\left(2.0 \times 10^{-2} \text{ m}\right)}{\pi \left(1 \times 10^{-3} \text{ m}\right)^4} = 2.0 \times 10^8 \frac{\text{Pa} \cdot \text{s}}{\text{m}^3}.$$

[Note that even though the combined cross-sectional area of the two parallel tubes is the same as the single tube, their equivalent combined resistance R_p is still *twice* that of the single tube R_1. This

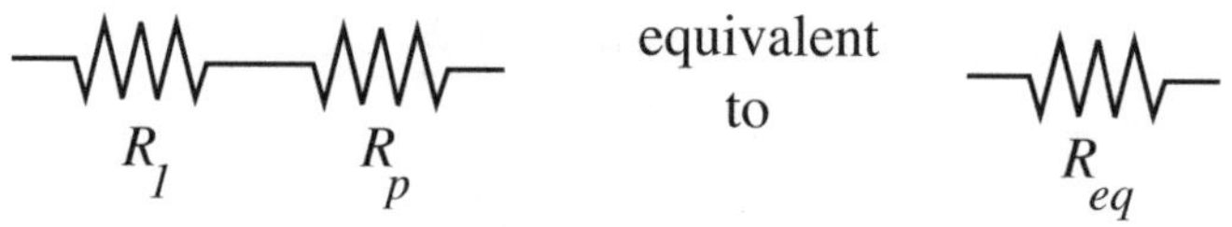

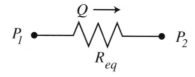

Figure 3.16: Two series resistances combined into one overall version.

shows the large influence that the r^4 term has in determining the resistance of an assembly, and why the arterioles can be so effective in controlling blood flow distribution by changing their diameter under smooth muscle control.]

Thus, $R_{eq} = R_1 + R_p = 2.0 \times 10^8 + 4.0 \times 10^8 = 6.0 \times 10^8$ Pa · s/m^3. The model is now simplified to a single resistance. See Fig. 3.17 where

$$Q = 1.0\frac{mL}{min} = 1.0 \times 10^{-3}\frac{L}{min}\left(\frac{1\ min}{60\ s}\right)\left(\frac{1.000 \times 10^{-3}\ m^3}{1\ L}\right) = 1.7 \times 10^{-8}\frac{m^3}{s}.$$

Figure 3.17: Branching tubes reduced to one overall resistance to find ΔP.

Then using (3.7), $\Delta P = Q \cdot R_{eq} = (1.7 \times 10^{-8}\ m^3/s) \cdot (6.0 \times 10^8\ Pa \cdot s/m^3) = 10$ Pa.

Converting to units of mmHg, $\Delta P = 10$ Pa $\cdot \left(\dfrac{1\ mmHg}{133.3\ Pa}\right) = \textbf{0.075 mmHg}$.

3.5   PROBLEMS

3.1. a. Given that an oxygen molecule in an aqueous environment will diffuse an average distance of 100 μm in 1.0 second, find its diffusion constant D from (3.1).

$$[\text{ans: } D = 2.5 \times 10^{-9}\ m^2/s]$$

b. Using the value from part **a**, estimate how long it would take an oxygen molecule to *diffuse* from your lungs to your big toe.

$$[\text{ans: Approx. } 1.8 \times 10^8\ s, \text{ or 5.8 yrs for a typical height!}]$$

3.2. Three tubes each have the same fluid resistance, $R = 6060$ Pa·s/m³ (to 4 significant figures).

   a. If all three are put in series, what is the equivalent overall resistance of the series combination?

$$[\text{ans: } R_{eq} = 18,180 \text{ Pa·s/m}^3]$$

   b. If all three are put in parallel, what is the equivalent overall resistance of the parallel combination? Hint: Use Equation (3.21).

$$[\text{ans: } R_{eq} = 2020 \text{ Pa·s/m}^3]$$

3.3. a. In the human cardiovascular system, the blood pressure measured at the arterial side (input side) of the capillary bed is approximately 6650 Pa, while the blood pressure at the venous side (output side) is approximately 3325 Pa. These values are for a typical human volumetric blood flow of 92 mL/s. (Note the use of mL here, a non-SI unit.) Using the concept of fluid resistance, calculate the total equivalent resistance of the capillary bed in SI units. Remember that $1L = 1.000 \times 10^{-3}$ m³ $= 1.000 \times 10^3$ cm³.

$$[\text{ans: } R = 3.6 \times 10^7 \text{ Pa} \cdot \text{s/m}^3]$$

   b. In the actual clinical environment, SI units are rarely used. In clinical practice, blood pressure is almost always measured in units of millimeters of mercury (mmHg), and blood flow is given in units of either liters per minute (L/min) or liters per second (L/s). In these units, the blood pressure across the capillary bed drops from 50 mmHg on one side to 25 mmHg on the other when the flow is 0.092 L/s. Using these *clinical* units, calculate the equivalent resistance of the capillary bed.

$$[\text{ans: } R = 270 \text{ mmHg·s/L}]$$

3.4. a. A single capillary is about 8.0 μm in diameter and has a typical length of about 1.0 mm. Using Poiseuille's Law and values for whole blood from Appendix B, calculate the fluid resistance (in SI units) of a *single* capillary tube.

$$[\text{ans: } R \approx 4.0 \times 10^{16} \text{ Pa} \cdot \text{s/m}^3 \text{ assuming a whole blood viscosity of 4.0 cP}]$$

   b. Of course, the human capillary bed is composed of many, many capillaries arranged in parallel. Assuming that each one has the resistance value found in part **a**, calculate how many of them must be in *parallel* in the human CV system such that the *overall* equivalent resistance has the value found in Problem 3.3a.

$$[\text{ans: about 1,100,000,000 !}]$$

c. In a sentence each, discuss how well this *capillary* configuration meets *each* of the requirements found on pp. 34 to 35 for Poiseuille flow to be valid:

1) The length of the tube must be much greater than the radius,

2) The flow must be steady in time and laminar in velocity profile,

3) The fluid must be Newtonian, and

4) The tube must be rigid.

CHAPTER 4

Hooke's Law: Elasticity of Tissues and Compliant Vessels

4.1 INTRODUCTION

Superman^{TM} may be the Man of Steel and his chest may stop bullets, but if his body were made entirely of steel-like material, it would not function nearly as well as yours or mine. For example, how could the skin around his elbow stretch and conform when it is bent, and how could the skin of his face show a smile? How could his bladder, if made of rigid material, expand to fill with urine, then contract as it is expelled? Without elasticity, his arteries and veins would not cushion and smooth out the pulses of blood flow from his heart, leading to punishing, pounding pressure waveforms throughout his body. And how could Superwoman^{TM} accommodate a pregnancy if her skin and abdominal organs could not expand or contract?

In fact, all tissues in the body have elasticity, some more so than others depending upon their function. Even bones, which are relatively stiff and rigid in order to act as the skeleton framework for the body, have some elasticity; otherwise they would not absorb shock and distribute stress appropriately. The soft tissues of the body are obviously more elastic than bone in performing their function. Cartilage is somewhere between bone and soft tissue in flexibility.

4.2 THE ACTION OF FORCES TO DEFORM TISSUE

There are three main classes of actions that any given external force can have on a tissue sample, depending upon the direction of the force and the direction of the distortion of the sample. These are summarized in Fig. 4.1, which considers a small stylized cube of the tissue.

In the first action, the force components F push perpendicularly inward equally on all faces of the sample, causing it to compress its volume. The magnitude of the force is best described in terms of the pressure $P = F/A$, where A is the face area upon which the force is acting. We have already encountered the concept of pressure in the previous two chapters.

In the second action, the force acts tangentially (sideways) to one or more faces of the sample, causing is to twist out of shape, but not changing its volume. The action of the force is best described here by the shear stress $\tau = F/A$, where A is again the area of the face over which the force is tangentially applied. We have seen how shear stress is related to fluid flow and viscosity in Chapter 3.

The third class of deformation is when the forces act only on two opposite ends of the sample, tending to pull it apart (or push it together). This puts the material in tension (or compression) and can best be related to the applied **tensile stress** $\sigma = F/A$, where again A is the cross-sectional area

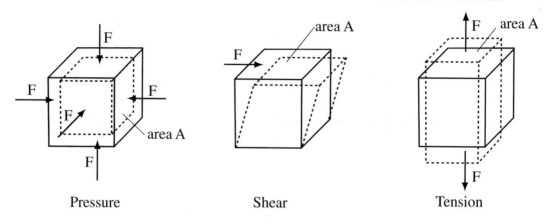

Pressure   Shear   Tension

Figure 4.1: Three ways that a force can distort a tissue sample.

of the face through with the force is applied perpendicularly. This is the deformation described by Hooke's Law below.

Note that the dimensions of all three stresses in Fig. 4.1—pressure, shear stress and tensile stress—are the same (force per area) and therefore their units are the same (Pa in SI units). However, how they are applied is different in each case. Pressure and tensile stress are both applied normally (i.e., perpendicularly) to the faces, while shear stress is applied tangentially. Pressure acts equally on all sides of the sample, while tensile stress is applied only to opposite faces.

4.3 HOOKE'S LAW AND ELASTIC TISSUES

Biomechanical engineers measure the extent to which tensile stress can distort various tissues by employing a testing machine similar to that shown in Fig. 4.2. A sample of the material to be tested (aligned such that the direction of stretch will be in a direction of interest in the sample material) is fabricated and clamped between a fixed base and a movable upper bar. The sample is usually fabricated in a "dog-bone" shape, the large ends for convenience in clamping and the narrower middle section to allow enough stretch for accurate measurement. Bones, ligaments, metals, man-made composites and some soft tissues can be tested in this manner.

When the test is started, the upper bar moves slowly upward (or downward for compression) in a controlled fashion, exerting a force on the sample. The magnitude of the force F is measured by a gauge (a load cell) in series with the sample.

Since the cross-sectional area A of the sample in its narrow region can be measured prior to the test (and is often assumed to remain approximately constant throughout the test[1]), the tensile **stress** σ at any time during the test can be calculated from

---

[1]When the cross-sectional area A is assumed to be constant, the stress is defined as "engineering" stress, as used here. "True" stress instead puts a changing area, which is more difficult to measure, in the denominator of (4.1).

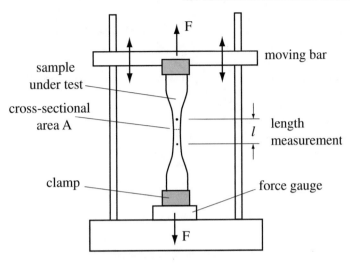

Figure 4.2: A tensile (or compressional) testing machine for determining the elastic constants of various materials.

$$\sigma = F/A. \qquad \text{(stress)} \tag{4.1}$$

The amount of stretch (or compression) of the sample during the test is measured by reference to two points separated by a distance l along the sample's axis. The measurement can be done with a number of different length gauges, or extensometers, including optical means and video cameras. With no applied force the points are a distance l_0 apart, the original spacing. As the force is gradually increased, their spacing changes to a new value l. The resulting **strain** ε of the material can be determined from[2]

$$\varepsilon = (l - l_0)/l_0. \qquad \text{(strain)} \tag{4.2}$$

Over the course of the test, values of both stress and corresponding strain are recorded. Note from (4.2) that strain is a dimensionless (therefore unitless) quantity, while from (4.1) stress has units appropriate for pressure. When plotted on a stress-strain curve, the results can appear as shown in Fig. 4.3 for two different materials.

The solid line in Fig. 4.3 plots the results for a material that behaves in a linear fashion, at least up to the point where the material fractures. For a linear elastic material, the strain and stress are related by a constant coefficient and obey Hooke's Law:

---

[2]When strain is defined with the original distance l_0 in the denominator, as in (4.2), it is known as "engineering" strain, as used here. "True" strain instead puts the changing distance l in the denominator, but this is a little more difficult to employ in most analyses.

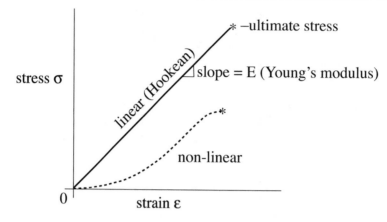

Figure 4.3: Results of a stress-strain test on two different materials. The material giving the solid line is linear with a slope equal to Young's modulus. The dashed line describes a non-linear material.

$$\boxed{\sigma = E\,\varepsilon.}\qquad \text{Hooke's Law} \qquad\qquad (4.3)$$

The constant of proportionality E in (4.3) is known as the **elastic modulus**, or **Young's modulus** or simply **stiffness**. It is found from the slope of the linear portion of the stress-strain plot. Since units of stress are the same as pressure (namely Pa) and strain is dimensionless, (4.3) shows that the units of E are also the same as pressure (Pa). Materials that follow the linear Hooke's Law are called Hookean. Bone and some biopolymers as well as biomaterials such as metal and ceramic orthopedic implants are approximately Hookean over a wide range of applied stress. Appendix B contains the Young's modulus for typical materials.

The dashed line in Fig. 4.3 shows a material that is highly non-linear in behavior, being pliable at low force, stiffening up for moderate to high force, then again getting pliable just before if fractures. This response is typical of many soft tissues in the body, such as bladder walls, skin and the walls of blood vessels. However, every tissue is somewhat unique in its exact response to force, depending upon its function. Many materials have a limited range over which their response is approximately linear, and they can be specified by a local tangent elastic modulus that is valid when the material is within that range.

As the force is applied larger and larger in Fig. 4.3, eventually all materials will fail by fracturing. The stress point at which this happens is called the **ultimate stress**, denoted by an asterisk (*) on the plot. It is the highest stress a material can tolerate without failure. Along with Young's modulus (if the material is linear), it characterizes the material. For example, with advancing age bone becomes more brittle and susceptible to fracture from falls. This is manifest in a lowering of its ultimate stress. The bone structure loses density and becomes osteoporotic. The risk for this condition is especially prevalent in post-menopausal women.

Example 4.1.   Bone Splint
When a splint (a bone support plate) is fixed along the side of a broken bone, it is important for the splint to be rigid enough to keep the bone at the fracture plane from moving under load, thus allowing the bone to stay in place and heal. Consider two different splint materials, polyethylene (a polymer) and stainless steel. Each splint has a cross section of 0.500×0.250 inches, and is 10.0 inches long. When a force of 200 pounds is applied to the top of the bone/splint combination, how much relative movement occurs between the two sides of the fracture for each material?

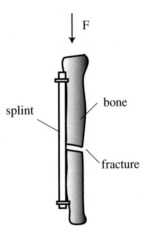

Figure 4.4: Sketch of bone splint configuration analyzed in Example 4.1.

Solution
Since the two faces of the bone at the fracture plane can slide and therefore will not resist any weight, all of the force will be absorbed by the splint. From (4.1) the tensile stress placed on the splint is

$$\sigma = F/A = 200 \text{ lbf}/(0.125 \text{ in}^2) = 1600 \text{ psi} = 1.10 \times 10^7 \text{ Pa}. \tag{4.4}$$

where a conversion factor from Appendix A was used to get the last value. The resulting strain is, from (4.3),

$$\varepsilon = \sigma/E. \tag{4.5}$$

Since by the definition of (4.2) $\varepsilon = (l - l_0)/l_0 = \Delta l/l_0$, solving for Δl and using (4.5) gives

$$\Delta l = l_0 \varepsilon = l_0 \sigma/E. \tag{4.6}$$

Now $l_0 = 10.0$ inches $= 0.254$ m, so using (4.4), $l_0\sigma = 2.79 \times 10^6$ Pa $\cdot$ m.
We will next put this value into (4.6) to solve for Δl for the two different materials with different Young's modulus.

Polyethylene: From Appendix B, Young's modulus for high-molecular-weight polyethylene is $E = 1.00 \times 10^9$ Pa. Therefore, the compression of the splint (which is seen as relative movement between the two sides of the bone at the fracture plane) is, from (4.6),

$$\Delta l = (2.79 \times 10^6 \text{ Pa} \cdot \text{m})/(1.00 \times 10^9 \text{ Pa}) = 2.79 \text{ mm.}$$

This large amount of movement will certainly be disruptive to the healing process.

Stainless steel: From Appendix B, Young's modulus for stainless steel is $E = 180 \times 10^9$ Pa. Thus,

$$\Delta l = (2.79 \times 10^6 \text{ Pa} \cdot \text{m})/(180 \times 10^9 \text{ Pa}) = \mathbf{15.5 \ \mu m.}$$

This movement is much smaller, and can be tolerated by the bone healing process.

4.4   COMPLIANT VESSELS

Tissue elasticity plays a major role in the proper functioning of the cardiovascular system. The vessels that carry blood from and to the heart are not rigid tubes, but rather have flexible walls that stretch in response to the blood pressure inside them, to a greater or lesser degree depending upon their elasticity and the pressure[3]. This vessel compliance has at least three important consequences on the nature of the circulation:

- The pulses of blood discharged by the heart ventricles with each beat are propelled into compliant vessels (the aorta and the pulmonary artery) that temporarily store some of the energy of the ejected blood as potential energy in their stretched walls. After the ventricle output valves close—blocking the return of the blood to the ventricles—the vessel walls recoil and the stored potential energy is converted to additional blood velocity (kinetic energy), helping push the blood down the arteries. This augments the cardiac output of the heart during the diastolic phase while at the same time keeping the peak systolic pressure lower. The actions of wall recoil are diagrammed in Fig. 4.5.

- The pulsating blood pressure, very noticeable in the aorta and major arteries, is damped by the compliance of the arteries, arterioles, and capillaries in conjunction with their resistance, so that by the time the blood reaches the capillaries, the pressure waveform is almost flat.

- The veins and venules are the most compliant of all the vessels, being thin-walled and very flexible, and they therefore store a good deal of the total blood volume (about 65%) of the

---

[3]Compliant blood vessels are called "windkessel" vessels, and are discussed in more detail in Chapter 5.

a. systole

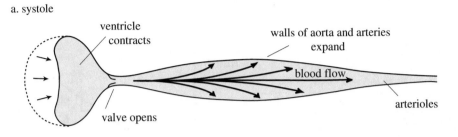

b. diastole

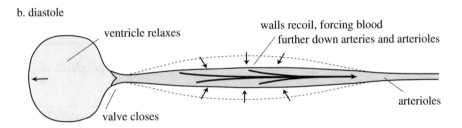

Figure 4.5: Diagram of the recoil action of the compliant walls of the aorta and arteries. The major effects of wall compliance are to lessen the peak pressure during systole (a), and to add flow down the circulation during diastole (b). (After Silverthorne, 1998.)

entire circulation. The size of this pool of blood is regulated by the compliance (the "tone") of the venous vessels, which in turn is under partial control by the nervous system. This forms an important contributor in the control of the blood distribution and pressure in the systemic system. In the case of a major change in the tone of venous compliance, such as happens in anaphylactic shock, there is a large change in the distribution of blood, leading to severe alteration of the circulation's effectiveness, and possibly even death.

A simple diagram of how the volume in a compliant vessel is related to the net pressure (i.e., inside pressure minus outside pressure) is shown in Fig. 4.6. Increasing the net pressure will linearly increase the vessel's volume—up to a point where the limit of elasticity is reached. In the linear region the proportionality constant between pressure P (Pa) and volume V (m^3) is called the vessel **compliance** C (m^3/Pa). The larger the vessel's compliance, the larger the volume for a given pressure.

Mathematically, the relationship shown in Fig. 4.6(b) can be stated as

$$\boxed{V = V_\phi + CP} \tag{4.7}$$

where P = net pressure in the vessel
  V = volume of the vessel

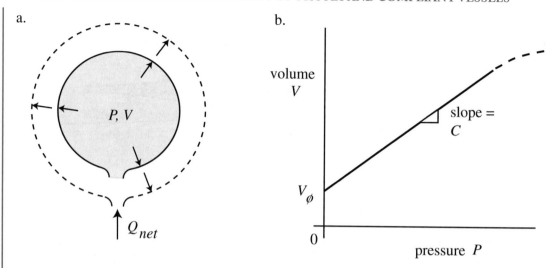

Figure 4.6: When the pressure increases in a compliant vessel, its volume increases (a). The graph of the volume vs. pressure (b) usually has a linear region, whose slope is the vessel compliance C.

C = compliance

V_ϕ = residual volume.

Equation (4.7) is a generalized form of Hooke's law applied to compliant vessels. The **residual volume** V_ϕ is the volume left in the vessel even with zero pressure; that is, it is the volume remaining when the vessel is completely relaxed. On the plot above, it is shown by the intercept of the straight line with the vertical axis. Arteries and arterioles have some residual volume (as evidenced by the fact that their diameters are still partially open when they are excised from the body), but venules and veins have much less residual volume and can collapse completely shut when unloaded.

The pressure P in (4.7) is the **transmural** ("across the walls") pressure, which is the difference between the pressures inside and outside of the vessel. Usually the pressure outside —which is the background pressure—is constant, and is often considered *zero* as the reference pressure.

Compliant vessels are found throughout the human body (blood vessels, bladders, lungs) as well as in other non-biological applications (hydraulic reservoirs, vibration dampers, automobile tires). Since they have the capacity for the storage of fluid volume, in diagrams they are given the symbol of a capacitor in Fig. 4.7.

This symbol has certain significance. The two horizontal lines can be thought of as representing the outline of a storage container (with variable volume). The fact that there is no direct connection between the upper line and lower line indicates that there is no net leakage across the walls from inside the vessel to the outside. (Leakage would be represented by a resistive element in parallel with the capacitor to ground.)

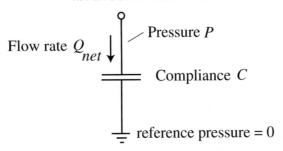

Figure 4.7: The symbol for a compliant vessel.

4.5   INCOMPRESSIBLE FLOW INTO AND OUT OF COMPLIANT VESSELS

To get fluid volume in and out of the vessels, there must be some fluid flow through one or more openings. There is a simple relationship between the volumetric flow rate Q (a variable introduced in previous chapters) and the volume V contained in the vessel. Since the liquids we are interested in (water and blood) are essentially incompressible, the volume of the fluid must be conserved when flowing from one region to another; this is the **principle of conservation of volume**. It means that during any increment of time Δt, a net volume flow rate Q into or out of the vessel will change the vessel's volume by the total amount of volume ΔV that has flowed, so

$$Q_{\text{net}} = \frac{\Delta V}{\Delta t}. \tag{4.8}$$

Rearranging,

$$\boxed{\Delta V = Q_{\text{net}} \Delta t.} \tag{4.9}$$

Assuming that the volume of the vessel originally had a value of V_{orig} before flow was measured and time incremented, then (by the definition of volume change) the new volume V_{new} after accounting for the flow Q_{net} is given by

$$V_{\text{new}} = V_{\text{orig}} + \Delta V = V_{\text{orig}} + Q_{\text{net}} \Delta t. \tag{4.10}$$

If more fluid is entering the vessel than is leaving during Δt, the net flow is in the direction *into* the vessel and the sign of Q_{net} is positive. Then (4.9) shows that the vessel volume change ΔV has a positive sign, and (4.10) shows in turn that the vessel volume increases. On the other hand, if more fluid *leaves* than enters, the net flow Q_{net} has a negative sign, ΔV has a negative sign, and the volume decreases.

Now from (4.7), pressure P and volume V are related by C, so the new pressure and new volume are related by

$$V_{\text{new}} = V_\phi + C P_{\text{new}}. \tag{4.11}$$

Rearranging,

$$P_{\text{new}} = \frac{V_{\text{new}} - V_\phi}{C}, \tag{4.12}$$

where V_ϕ is the residual volume of the vessel and shouldn't be confused with V_{orig}. The value for V_{new} is obtained from (4.10). These relationships are important in understanding the pressure/volume interaction in vessels of the human cardiovascular system.

Extending (4.7) in another way, since pressure P and volume V are related by C, then *changes* in pressure and volume are similarly related. From (4.7), since the residual volume is a constant *and assuming C is a constant*[4],

$$\Delta P = \frac{\Delta V}{C}. \tag{4.13}$$

Substituting ΔV from (4.9) into (4.13) and rearranging gives

$$\frac{\Delta P}{\Delta t} = \frac{1}{C} Q_{\text{net}}. \tag{4.14}$$

For (4.14) to be accurate, Q_{net} and C must be steady during the interval Δt. If Δt is small enough, this is true. In fact, taking Δt to a limit approaching zero gives the differential form of (4.14):

$$\frac{dP}{dt} = \frac{1}{C} Q_{\text{net}}. \tag{4.15}$$

Equations (4.14) and (4.15) state that the more compliant a vessel is (i.e., the larger C is), the slower the pressure will change for a given flow rate in or out.

Example 4.2. Capillary Compliance
The human systemic capillary bed has a mean blood pressure of about 30 mmHg and holds about 6.0% of the total body blood volume. Its residual volume is about 1.0% of the total blood volume. Estimate its total compliance.

Solution
Since the total blood volume in a human is about 5.0 L, the systemic capillary bed has a residual volume $V_\phi = 0.010 \times 5.0 \, \text{L} = 0.050 \, \text{L}$. Its total volume at a pressure of 30 mmHg is $V = 0.060 \times 5.0 \, \text{L} = 0.30 \, \text{L}$. Thus, from (4.7)

---

[4]The compliance of most vessels can be considered constant over a short time period, but NOT that of the heart ventricles (see Chapter 5).

$$C = (V - V_\phi)/P = (0.30 \text{ L} - 0.050 \text{ L})/30 \text{ mmHg} = \mathbf{0.0083 \text{ L/mmHg}}.$$

Example 4.3.   Pressure Increase Caused by Blood Flow
If during a period of 5 seconds, the blood flow into the capillary bed of the previous example is 5.6 L/min, but the flow out is 5.2 L/min, how much would the capillary volume and blood pressure change during this period?

Solution
The net flow into the capillary bed is given by the inflow minus the outflow, or
$Q_{net} = (5.6 - 5.2)$ L/min $= +0.4$ L/min $= +0.007$ L/s. According to (4.9),

$$\Delta V = Q_{net} \Delta t = (0.007 \text{ L/s})(5 \text{ s}) = +0.04 \text{ L}.$$

From (4.10) and the values of the previous example,

$$V_{new} = 0.30 \text{ L} + 0.04 \text{ L} = 0.34 \text{ L}$$

and from (4.12),

$$P_{new} = (0.34 \text{ L} - 0.05 \text{ L})/0.0083 \text{ L/mmHg} = \mathbf{35 \text{ mmHg}}.$$

So the capillary blood pressure would rise from 30 mmHg to 35 mmHg.

4.6   PROBLEMS

4.1. m.[5] Suppose that the following lines are typed in the command window of Matlab. After each line is typed, the "return" or "enter" key is hit. First, on a piece of paper *without* the aid of a computer, write down what you believe would be printed or plotted on the Matlab screen in response to each line. Then check your answers by actually entering these lines, one at a time, into the command window of a computer running Matlab. Correct your initial answers on the sheet of paper. (Notes: For this homework, just turn in your hand-written final answers; you don't need to print out the Matlab screen. Also, if you don't understand why you are getting certain responses, try using the help command or the help desk feature of Matlab.)

```
A=ones(1,5)
a=[5 6 4 2 1]
a(3)
```

[5]Problems with the suffix "m" have significant Matlab content.

```
a(3)=0
b=[1:5]
b(4:5)
c=a + b;    %careful, this one is tricky
d=a + A
pause(4)
```

4.2. m. Suppose that the following lines have been stored as an m-file under the name "work2.m." First, on a piece of paper *without* the aid of a computer, write down what you believe would be printed or plotted on the Matlab screen in response to typing `work2` in the command window of Matlab. Then check your answers by actually storing these lines as an m-file with the name "work2.m," and then typing `work2` in the command window. (Make sure your working Matlab *path* includes the directory that has this m-file inside so Matlab can find it. You can use a menu command to add the directory to Matlab's path if necessary.) Correct your initial answers on the sheet of paper. (Note: Just turn in your hand-written answers along with a rough sketch of the plot on your sheet; you don't need to print out the Matlab screen.)

```
f=[6 3 2 8; 5 4 2 1;
     9 0 0 3]
f(3,1)
f(2,:)
f'
clear f
f=[3:2:12]
f(4)=f(3)
g=[1 2,3 4 5]
plot(f);  hold on;  plot(g)
xlabel('index number')
ylabel('f and g')
text(3,5,'great plot!')
```

4.3. m. Write a Matlab m-file (called a "script") that does the following tasks in order:

a. Set up a row vector that has 21 elements ranging linearly in value from 0 to π.

b. Using the vector from part **a**, produce a row vector that has the shape of the positive half-cycle of a sine wave. It will have values ranging from 0 to 1.

c. Using the vector from part **b**, produce a row vector that has the same shape as part **b**, but has values ranging from 0 to 10.

d. Using a `for loop`, search each element of the vector from part **c** to test whether the element has a value less than 7.0. If it does, set the element value to 0; if it doesn't, set the value to 12.0. (Note: use an if statement inside the `for loop`).

e. Add up the values of all the elements in the vector as modified in part **d**. Write out this value to the screen.

Write this Matlab program by hand first. Then type it in as an m-file, store it, and run it with Matlab. Print out a copy of your m-file only and turn it in along with the answer to part **e**. (Note: Do not print out or turn in your element values for the row vector. Also, if you don't have access to a printer, turn in a hand-written copy of the program.)

[ans: 132]

4.4. **m.** On a clean piece of paper, write down what would be displayed ("printed") on a computer screen in response to each of the lines below when they are typed one at a time (followed by an enter) in the command window of Matlab. (You don't need to repeat the commands on the paper.) Write down *only* what would be displayed; do all other calculations you might need on another piece of paper that you do not turn in.

```
a = ones(1,6)
b = [1:2:11]
b(3) = 0
c = a+b
d = 2*c
for i = 1:6;
if d(i) > 8; d(i) = 8;
else d(i) = 0;
end;
end;
d
```

4.5. Approximate your tibia as a long cylinder bone. Measure its approximate length and estimate its diameter. When you stand on one leg, how much total shortening occurs in the length of your tibia? (See Appendix B for the Young's modulus of bone.)

[ans for me: about 5.6 μm]

4.6. **a.** The pulmonary artery contains about 52 mL of blood (averaged over time) at an average pressure of 17 mmHg. Its residual volume is 21 mL. Calculate its compliance.

[ans:$C \approx 0.0018$ L/mmHg]

b. At the end of diastole (thus the beginning of systole), the pressure in the pulmonary artery has dropped to about 10 mmHg. How much blood is in the artery at that point in time?

[ans:$V \approx 39$ mL]

c. During the first 0.10 s of systole, the flow rate of blood from the right ventricle into the pulmonary artery is 0.24 L/s, while the flow rate out of the artery toward the lungs is 0.15 L/s. At the end of this 0.10 s interval, what is the blood pressure in the artery?

[ans:$P \approx 15$ mmHg]

4.7. The *heart ventricles* are compliant vessels, but they have one very important difference from ordinary blood vessels: their compliance *changes* dramatically during each heart cycle, from a relatively large value during filling (diastole) to a small value during ejection (systole).

a. The end-diastolic volume of blood, V_{ED}, at the very end of diastole in the left ventricle is about 135 mL at a filling pressure of 8.0 mmHg. What is its end-diastolic compliance? (Assume the residual volume of the left ventricle is negligible.)

[ans: $C = 0.017$ L/mmHg]

b. At the peak of systole—which occurs about 175 ms after the end of diastole—the left ventricle compliance has a value of 0.000800 L/mmHg. The outflow of blood into the aorta averages 200 mL/s during systole. What is the systolic pressure in the ventricle at this moment?

[ans: $P = 125$ mmHg]

4.8. Tissues that are composed of randomly oriented cross-linked collagen fibers are in some ways analogous to rubber. The elasticity of these types of tissues and materials can be measured using a standard tension test. Your homework is to perform such a test at home or in the lab. A schematic of the experimental set-up is shown below in Fig. 4.8.

a. Put two ink spots on a moderately thick rubber band and then tie one end to an empty soda can and the other end to a strong horizontal rod (or tree limb…). Slowly fill the soda can with water and measure the length between the two ink spots for various amounts of water. Each data point will consist of two corresponding values: the length between the spots and the weight of the water in the can. Using Matlab, enter the length values as one row vector, and the corresponding weight values as another row vector (of the same length). Then, using Matlab, plot the applied force (weight of water) on the y-axis versus the length on the x-axis. Label each axis.

Note: The force due to the weight of the water is $F = \rho g V$, where density $\rho = 999$ kg/m^3, the acceleration of gravity $g = 9.8067$ m/s^2, and $V =$ volume of water in m^3.

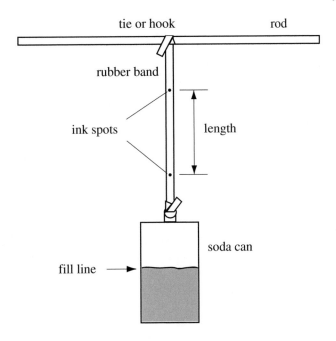

Figure 4.8: Experimental set-up for determining Young's modulus of a rubber band.

b. Then convert the data to two new row vectors, one representing stress (force divided by the initial cross-sectional area of the rubber band) and the other representing strain (*change* in length divided by initial length). Plot stress vs. strain (similar to Fig. 4.3), labeling the axes.

c. Finally, calculate in SI units the elastic modulus (Young's modulus) E of the rubber-band material. Is the material Hookean, i.e., linear elastic? Please staple your rubber band (or a short piece of it) to your answer sheet, and turn in your printed Matlab plots. If you don't have access to a printer, sketch the plots by hand.

> [ans: for a typical rubber band: $E \approx 0.7$ MPa, but it may vary considerably depending upon the particular rubber band and the range selected. The rubber band is not strictly Hookean, but can be approximated as linear over some range.]

CHAPTER 5

Starling's Law of the Heart, Windkessel Elements and Conservation of Volume

5.1 INTRODUCTION – COMPLIANCE OF THE VENTRICLES

As discussed in Chapter 4, all blood vessels in the human body have compliance, which means that they expand with increasing blood pressure inside. (Actually, all tissues of the body, not just blood vessels, have various degrees of elasticity, even the relatively stiff bones and teeth.) The blood vessels on the venous side of the circulation have the most compliance, and the arterial vessels have less. As mentioned in the previous chapter, this compliance helps smooth out the pulsatile nature of blood flow, keeping the flow steadier while reducing peak pressures, and provides a controllable storage volume for the blood, especially in the veins.

The walls of the chambers of the heart (the two atria and two ventricles) also exhibit compliant behavior. Indeed, as they are receiving considerable blood inflow each heart cycle during their **diastolic** (relaxing and filling) phase, they must expand easily to accept this volume. But, after filling, they then must contract forcefully to start the **systolic** (contraction and ejection) phase of the heart cycle. Thus, the heart wall muscles change from *high* compliance (low stiffness) during relaxation to *low* compliance (high stiffness) during contraction, then back again, cycle after cycle.

The left ventricle plays an important role in the heart, so let's focus on its action. At the very beginning of systole, the volume of blood that filled the ventricle during diastole has not yet been expelled into the aorta through the aortic valve. Therefore, the ventricular volume at the start of systole is the same as at the end of diastole. But the force exerted on this blood by the ventricular walls (measured by pressure inside the ventricle) sharply rises, marking the beginning of the systolic period. This phase is termed **isovolumic contraction**, because the contraction force increases dramatically without any change (at least at first) in ventricular volume. A symmetrical situation occurs at the end of systole, which is the beginning of diastole, when the ventricle relaxes. This phase is termed **isovolumic relaxation.**

One revealing way to characterize these events is to quantify the changing compliance of the left ventricle. During diastole, when it is filling by accepting blood through the mitral valve, the left ventricular walls are relaxed and therefore are characterized by a relatively large compliance C_{hd}. During systole, when the muscular walls stiffen to force blood out, they are characterized by a much smaller compliance C_{hs}. Then, at the end of systole, the walls relax again and the ventricle returns to the more compliant state for the next filling phase. A good approximation of this alternating compliance is shown in the graph of Fig. 5.1.

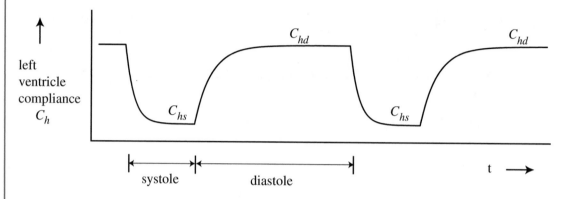

Figure 5.1: The approximate variation of left ventricular compliance during a heart cycle.

What happens to blood pressure as a result of this alternating compliance? As Equation (4.7) in the Hooke's Law chapter showed, pressure and volume are related in general by the compliance of the vessel:

$$V = V_\phi + CP, \tag{5.1}$$

so pressure P depends upon both the volume V (which itself is changing during the heart cycle) *and* compliance C of the ventricle. (V_ϕ is the residual volume of the vessel—for dynamic modeling of the left ventricle, this is considered to be small.)

Rearranging (5.1) to solve for the pressure in the left ventricle (and specializing the variables here by the subscript h, denoting the heart),

$$P_h = (V_h - V_\phi)/C_h. \tag{5.2}$$

Thus, during systole when the left ventricle compliance C_h drops rapidly to a low value ($C_{hs} \ll C_{hd}$), since the compliance value C_h is in the denominator of (5.2), the blood pressure in the ventricle is shown to rapidly rise. This increased pressure is the force that causes ejection of the blood out of the ventricle, leading to the pumping action of the heart.

5.2   PRESSURE-VOLUME PLOTS: THE PV LOOP

An illustrative way of viewing the changing pressure and volume of the left ventricle is to track its pressure as a function of volume throughout one heart cycle; the resulting graph will form a closed curve (a **PV loop**), as shown in Fig. 5.2. (This graph is, unfortunately, not trivial to obtain in humans, since it requires an invasive pressure probe placed inside the left ventricle, along with some means of measuring the changing ventricular blood volume, usually done with ultrasound imaging.)

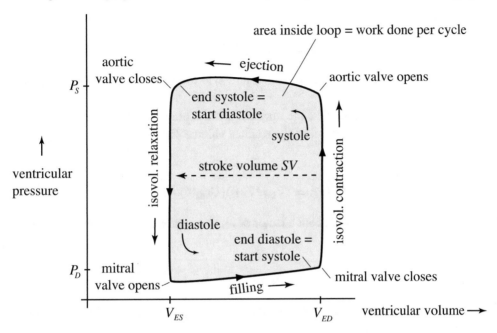

Figure 5.2: Pressure-volume graph (PV loop) of the pressure and volume changes in the left ventricle during one heart cycle. The loop progresses counterclockwise from the start of systole (lower-right corner) through the completion of systole and diastole before returning back to start the next cycle.

During each heart beat, the behavior of the ventricle progresses around the loop in a counterclockwise fashion. At the start of systole (end of diastole) in the lower-right corner, the ventricle rapidly stiffens (lowering its compliance to C_{hs}). The resultant increase in pressure causes the inlet mitral valve to close so the volume of blood is temporarily trapped inside the stiffening ventricle (the isovolumic contraction phase). When the ventricular pressure rises above the aortic pressure, the outflow aortic valve opens, allowing rapid ejection of blood and partially emptying the ventricle. If we label the pressure at the end of systole as P_S, (5.2) gives

$$P_S = (V_{ES} - V_\phi)/C_{hs}, \tag{5.3}$$

where V_{ES} is the volume of the heart at the end of systole (same as at the start of diastole). Since C_{hs} is low, P_S is high.

Then the ventricle relaxes (increasing its compliance to C_{hd}) marking the end of systole and the start of diastole. The falling pressure causes the aortic valve to close, again trapping the (smaller) blood volume in the relaxing ventricle (the isovolumic relaxation phase). When the pressure drops below the venous pressure, the mitral valve opens, allowing low-pressure filling of the relaxed ventricle, setting the stage for the next contraction. If we label the pressure at the end of diastole as P_D, (5.2) gives

$$P_D = (V_{ED} - V_\phi)/C_{hd}, \tag{5.4}$$

where V_{ED} is the volume at the end of diastole (same as at the start of systole). Since C_{hd} is high, P_D is low.

It's straightforward to show that the nominal systolic pressure P_S is much higher than the nominal diastolic pressure P_D. Assuming the residual volume V_ϕ can be neglected here, the ratio of (5.3) to (5.4) gives

$$P_S/P_D = (V_{ES}/V_{ED})(C_{hd}/C_{hs}). \tag{5.5}$$

While V_{ES} is smaller than V_{ED} (by about a factor of 2), C_{hd} is much greater than C_{hs} (at least by a factor of 20), so (5.5) shows that

$$P_S \gg P_D, \tag{5.6}$$

which is a requirement for pumping the blood into the higher pressure arteries after being collected from the lower pressure veins.

Several other concepts can be gleaned from the PV loop in Fig. 5.2. The width of the loop between the volume at the right boundary (V_{ED}) and the volume at the left boundary (V_{ES}) is exactly the amount of blood pumped out during that heart cycle (equal to $V_{ED} - V_{ES}$). This is called the **stroke volume**, or SV, usually given in units of liters. Also, a consideration of the energy spent by the heart in pumping out this stroke volume of blood against the aortic pressure (to be covered in Sections 7.4 and 7.5) shows that the energy expended (i.e., the work) per heart beat is given by the **area** enclosed inside the loop; the bigger the loop area, the more energy is used.

If we now envision the passage of time during the heart cycle, we can plot the ventricular pressure and volume as a function of time. The graph would be similar to Fig. 5.3, which is close to what's observed in the normal human heart.

5.3   STARLING'S LAW OF THE HEART

The human body must be able to adapt to a wide variety of conditions and still maintain proper functioning. This general principle applies to the regulation of the volume of blood pumped by the heart. There are times when the degree of filling of the ventricle with returned blood is markedly higher than normal for a short (or even moderate) period. For example, a noticeable increase in filling volume often accompanies exercise, when muscular action forces more blood from the venous system into the right side (then into the left) of the heart. Similarly, when a person sneezes or forcefully

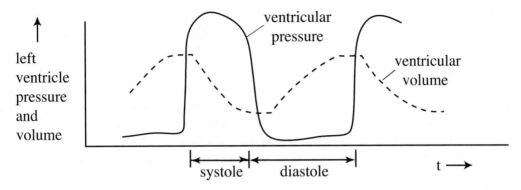

Figure 5.3: Pressure and volume waveforms for the left ventricle.

exhales a deep breath, the pressure inside his/her thorax temporarily increases, squeezing blood out of the veins and vena cava, filling the heart more fully[1].

Starling's Law of the Heart (also known as the Frank-Starling mechanism) states that the heart accounts for this increased filling volume by increasing its strength of ventricular contraction, thereby increasing cardiac output. Otherwise, blood would dam up in the heart and severely hamper its functioning. In Ernest Starling's words (1918):

> "If a man starts to run, his muscular movements pump more blood into the heart. As a result the heart is overfilled. Its volume, both in systole and diastole, enlarge progressively until by the lengthening of the muscle fibres so much more active surfaces are brought into play within the fibres that the energy of the contraction becomes sufficient to drive on into the aorta during each systole the largely increased volume of blood entering the heart from the veins during diastole…. The physiological condition of the heart is thereby improved and the heart gradually returns to its normal volume even though it is doing increased work."

In shorter but less elegant words: *The heart will pump out all of the blood that is returned to it, within the limits of the heart.*

The physiological mechanisms that underlie the Frank-Starling mechanism are well known, and include the fact that cardiac muscle fibers increase their contractility (contraction force) the more they are stretched, up to a point, and the fact that increased permeability to calcium ions in muscle cells, enhanced by stretching, also increases muscle contractility. We will not pursue these factors further here, but are more interested in the following question pertaining to modeling: Will modeling the ventricle as a vessel whose compliance changes between diastole and systole produce a behavior that is consistent with Starling's Law?

[1] Doing this exhaling action while purposely closing the glottis to block the air and thus accentuate the increase in intrathoracic pressure is called a "Valsalva maneuver." The heart rate also temporarily increases in response.

To answer that question, consider (5.2), the formula for systolic pressure in the ventricle. Note that as the volume of the ventricle increases, the systolic pressure increases in response. [Although (5.2) is written in terms of end-systolic volume, the concept also holds for increased end-diastolic (filling) volume.] Thus, systolic pressure increases as a consequence of increased filling, and volumetric flow into the aorta and cardiac output increases, consistent with Starling's Law. This increased cardiac output will continue (slowly declining) until the diastolic volume returns to normal.

An instructive way of viewing Starling's Law in action is to plot a succession of PV loops starting with a situation in which the left ventricle is overfilled compared to normal, then following the loops over the next several cycles as the heart returns to normal volume. Such a sequence is shown in Fig. 5.4, where the loop-to-loop progression is right to left in the figure. Note how the stroke volume is large for the starting loop to help rid the ventricle of excess volume, then decreases with each beat. Also the work done per beat (the area inside each loop) is large at the start, then decreases toward a steady-state value.

The line which marks the left edge of the end-systolic point (upper-left corner) of each successive loop has a slope inversely proportional to the minimum compliance (maximum stiffness) of the ventricle. Therefore, its slope is equal [2] to $1/C_{hs}$.

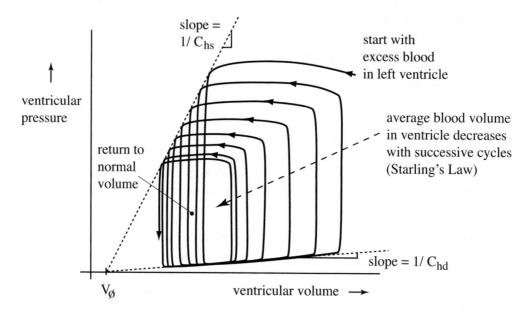

Figure 5.4: A sequence of PV loops starting with an abnormally filled ventricle. After several beats, the volume inside the ventricle returns to normal, illustrating Starling's Law of the Heart.

[2]This assumes that compliance of the ventricle reaches a value of C_{hs} at the end of systole, which is approximately true. Also, some researchers use a parameter named "elastance" instead of compliance to characterize the ventricle's stiffness, where elastance = 1/compliance. Thus, the slope of this upper line is also equal to maximum elastance.

Example 5.1.   Systolic Pressure
The pressure in the left ventricle of a heart model is about 10 mmHg near the end of diastole. The diastolic compliance of this particular ventricle is $C_{hd} = 0.013$ L/mmHg. At the beginning of systole, the compliance drops to $C_{hs} = 0.00050$ L/mmHg. Find the volume in the ventricle (minus the residual volume) at the start of systole, and estimate the systolic pressure assuming a stroke volume of 70 mL.

Solution
The beginning systolic volume is the same as the end-diastolic volume, found from (5.4):

$$(V_{ED} - V_\phi) = P_D C_{hd} = (10 \text{ mmHg})(0.013 \text{ L/mmHg}) = 0.13 \text{ L}.$$

At the end of systole, the ventricular volume has been reduced to $0.13 - 0.07 = 0.06$ L, and from (5.3) the systolic pressure is

$$P_S = (0.06 \text{ L})/(0.00050 \text{ L/mmHg}) = \textbf{120 mmHg}.$$

5.4   WINDKESSEL ELEMENTS

We have seen from Poiseuille's Law how fluid tubes have resistance to flow, and from Hooke's Law how blood vessels can be represented as compliant vessels. It is now time to put these two concepts together, because a realistic model of blood vessels should include *both* effects. Using symbols, a distributed model for the vessel will look like Fig. 5.5.

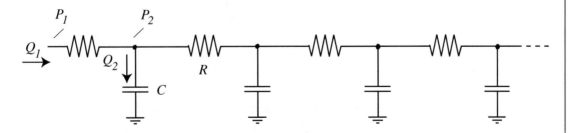

Figure 5.5: A distributed-element model of a blood vessel.

A blood vessel possessing both distributed resistance and compliance properties is called a **windkessel** element. The name comes from a German word for the compliant bellows attachment used with early fire engines to smooth out pulsatile flow of water from the engine. It is appropriate here, since as we have seen from the Hooke's Law unit, the windkessel action of the aorta and arteries tends to smooth out the pulsing nature of the flow from the ventricle due to the expansion, then recoil, of the walls with the pulsing pressure wave.

An accurate model of a windkessel element would include numerous resistive and compliant elements (perhaps with nonlinear properties) distributed along the length of the vessel, as in Fig. 5.5, whose values vary to represent the changing properties of the vessel along its length. However, it is much more convenient to model the vessel with just a few (even one or two) of each of the elements, and to approximate these as linear. Thus, a major vessel might be modeled as shown in Fig. 5.6.

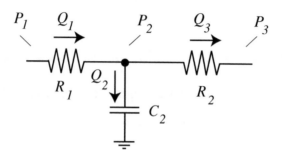

Figure 5.6: Simplified windkessel model of a blood vessel.

5.5   CONSERVATION OF VOLUME IN INCOMPRESSIBLE FLUIDS

Engineers like to know the value of quantities. That is, in addition to the important goal of understanding the concepts behind natural effects (a *qualitative* understanding), they want to be able to calculate and measure the magnitudes of the quantities involved (a *quantitative* understanding). For example, it is vital to keep track of the mass of a fluid as it flows from one region to another, such as the flow in the three branches in Fig. 5.6 above. Because the total mass of the fluid is neither being created nor destroyed in this situation, just moved, mass must be conserved as it flows. This is the **principle of conservation of mass**.

In addition, if the fluid is incompressible (or nearly incompressible as with water and blood), its density doesn't change. This in turn means that its total volume doesn't change under flow. This is the **principle of conservation of volume**, first introduced in Section 4.5. If the total volume is conserved, then any volume that leaves one region must add to the volume in neighboring regions. This leads to the conclusion that the rate at which volume leaves one region must be equal to the rate at which it enters other regions. Specifically, in Fig. 5.6 at the junction where P_2 is measured, the volumetric flow rate *in* is Q_1 while the flow rate *out* to other regions is Q_2 plus Q_3. Therefore, conservation of volume requires that

$$Q_1 = Q_2 + Q_3. \tag{5.7}$$

In words: *At a junction, flow rate in = flow rate out.* This principle can be applied to the general case of several branches converging on a junction by letting the outward flow terms have a negative sign and the inward flow terms have a positive sign. Then in general at any junction with N branches,

$$\sum_{i=1}^{N} Q_i = 0. \tag{5.8}$$

Example 5.2.   Conservation of Volume
In the circuit shown below, let the flow rate $Q_1 = 0.1$ L/s and $Q_4 = -0.5$ L/s. Also $P_2 = 50$ mmHg, $P_3 = 20$ mmHg, and $R_2 = 100$ mmHg $\cdot$ s/L . Find the flow rate Q_2.

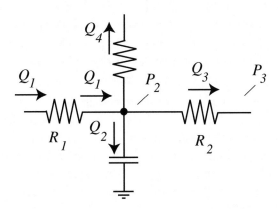

Figure 5.7: Schematic used to demonstrate conservation of volume in Example 5.2.

Solution
Since no volume can be stored in a resistance element, the flow rate Q_1 is constant through element R_1, and Q_1 enters from the left into the junction where P_2 is measured. Now from Poiseuille's Law:

$$Q_3 = (P_2 - P_3)/R_2 = (50 \text{ mmHg} - 20 \text{ mmHg})/100 \text{ mmHg} \cdot \text{s/L} = 0.3 \text{ L/s}.$$

Then, from (5.8),

$$Q_1 - Q_2 - Q_3 - Q_4 = 0,$$

or $Q_2 = Q_1 - Q_3 - Q_4 = [+0.1 - 0.3 - (-0.5)]$ L/s $= +0.3$L/s.

5.6   PROBLEMS

5.1. If you are doing the Matlab portion of the Major Project, you should start calculating the various R's and C's needed for the project following the procedure discussed in class. All calculations must be done from the very beginning in ink in your Major Project lab notebook. For this problem, copy *from your lab notebook* onto a separate sheet of paper (in pencil) the particular pressures, volume, and flow rate you used to find C_o and R_{oa} (the compliance of the aorta and the resistance between the aorta and arteries/arterioles) along with *your* calculated C_o and R_{oa} values. Turn in this sheet only.

> [my values: $R_{oa} = 141$ mmHg · s/L and $C_o = 0.00065$ L/mmHg; your values should be within a factor of about 2 of these.]

5.2. m. For the Major Project, you will need to generate a Matlab row vector that contains the curve of the changing compliance of the left ventricle, C_h. It should look like Fig. 5.1 or Fig. 3 in the Major Project. The goal of this problem is to generate a *single* appropriate row vector for C_h. The formulas of the compliance for the periods of systole and diastole are given in Eqs. (1) and (2) following Fig. 3 in the Major Project.

a. Write a script m-file to generate the proper row vector for C_h following the times shown in Fig. 5 in the Major Project. You will first have to use Eqs. (1) and (2) to find the values of C_h during systole and diastole, respectively, then use "cut and paste" to put these values into *one* vector which follows the timing shown in Fig. 5. Break time into 1-ms increments, so the vector is 800 elements long (corresponding to 0-799 ms). Have systole last from 100 ms to 349 ms and have diastole occupy the remaining segments of the cycle (from 0 to 99 ms, and from 350 to 799 ms). Let the time constants be $\tau_s = 30$ ms and $\tau_d = 60$ ms. Also, *for this problem only*, let the limits of compliance be $C_{hs} = 0.001$ and $C_{hd} = 0.01$ L/mmHg (these limits of compliance will be different in your actual Major Project). Read in the values of τ_s, τ_d, C_{hs}, and C_{hd} from a parameter file. Print a copy of both m-files.

b. Plot and print a curve of your vector C_h versus time. (Save your Matlab code for later use.)

> [Note: all you need to turn in for this problem are printouts of your two Matlab m-files from part **a**, and your plot from part **b**, shown in Fig. 5.8]

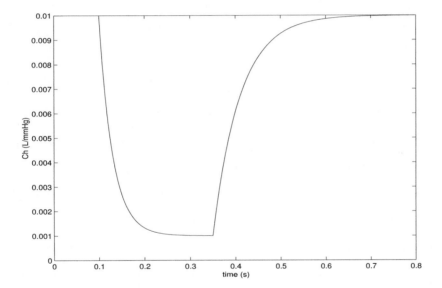

Figure 5.8: Matlab plot of changing compliance of left ventricle.

CHAPTER 6

Euler's Method and First-Order Time Constants

6.1 INTRODUCTION – DIFFERENTIAL EQUATIONS

We saw in Section 4.5 that the change in a vessel's volume ΔV during an interval of time Δt is related to the net flow of fluid Q (inflow minus outflow) into the vessel by the following equation:

$$\frac{\Delta V}{\Delta t} = Q. \tag{6.1}$$

This relationship assumes that the flow Q is constant during the time interval Δt. The assumption of constant Q becomes more accurate as the time interval Δt becomes smaller, since Q is less likely to change during a shorter interval. Taking the limit of the left-hand side of (6.1) as Δt goes to zero gives the differential form of (6.1):

$$\frac{dV}{dt} = Q, \tag{6.2}$$

where dV/dt is the time rate of change of V. This equation is an example of a first-order (i.e., one level of differentiation) differential equation, which in general can be written in the form

$$\frac{dV}{dt} = f(V), \tag{6.3}$$

where $f(V)$ is some function of the variable V.

There are two approaches to finding the solution to an equation such as (6.3). If $f(V)$ belongs to one of a number of classes of well-behaved functions, it may be possible to find an explicit closed-form solution for the variable V. This is known as an **analytic** solution to (6.3); an example of this will be seen later in this chapter.

But it is not always possible (in fact, is somewhat unusual in the real world) to find an analytic solution to a differential equation which describes a complex problem. In this case, an approach that will give an approximate, but arbitrarily accurate, solution to the differential equation is to use a digital computer to arrive at the answer. This approach, of which there are many versions, is called the numerical analysis approach, or **numerical** solution for short. Euler's method, discussed next, is a simple numerical analysis algorithm that works on many problems.

6.2 EULER'S METHOD

The derivative dV/dt is the limiting form of the ratio $\Delta V/\Delta t$ as Δt gets infinitesimally small. If we don't go all the way to the limit of Δt being zero, but rather stop at a small enough value of Δt that the local change in V, ΔV, is approximately linear, then we have the **finite-difference** form of (6.3), which is a good approximation of (6.3):

$$\frac{\Delta V}{\Delta t} = f(V) \tag{6.4}$$

Figure 6.1 shows ΔV and Δt at one particular segment along the curve of V versus t. The increment Δt must be small enough that the curve of V is approximately straight within the neighborhood of Δt. (Note that if V varies rapidly with t, Δt may need to be made smaller to make sure that the line segment of V is approximately straight within the extent of Δt.)

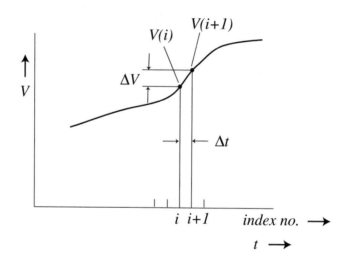

Figure 6.1: Finite-difference approximation to the derivative dV/dt, showing ΔV and Δt.

The curves of variables such as the volume V are often described in sampled form, where the values of V at uniform intervals are found and stored. This format, called the "vectorized" form of V, is handy for manipulation and storage by a digital computer. **Euler's method** is based upon relating successive time samples of V to each other. Specifically, as seen in Fig. 6.1 at each increment,

$$\Delta V = V(i+1) - V(i). \tag{6.5}$$

So (6.4) can be written

$$\frac{V(i+1) - V(i)}{\Delta t} = f(V(i)). \tag{6.6}$$

Rearranging, $\boxed{V(i+1) = V(i) + \Delta t \cdot f(V(i)).}$ Euler's Method   (6.7)

Thus, at each sample point with index i, the value of the *next* sample point $V(i+1)$ can be obtained from the present value $V(i)$. By starting out with some known initial value of V, later values of V can be successively calculated from (6.7). The solution for V is therefore obtained by stepping through the equations in time, with time increasing by Δt for each step.

Euler's algorithm (6.7) predicts the future value of a variable based upon the present value of the variable, and is therefore termed a *forward-difference* method[1]. In general, using a finite-difference numerical approach by stepping through differential equations in time is known as the *finite-difference time-domain*, or FDTD, technique for solving the equations.

Euler's method can be applied to (6.2), the equation describing the change in the volume in a vessel. In finite-difference form, (6.2) becomes

$$\boxed{V(i+1) = V(i) + \Delta t \cdot Q(i).}$$ Euler's Method   (6.8)

Remember that $Q(i)$ will usually be related to $V(i)$ through other equations which must be considered along with (6.8). An example of this technique is found at the end of this chapter.

6.3   WAVEFORMS OF PRESSURE AND VOLUME

The finite-difference forms of the equations for volume and pressure reveal a concept that will prove useful when analyzing the time behavior of these quantities. In particular, how much can the volume V "jump" between the two samples at (i) and $(i+1)$ in Fig. 6.1? As Δt gets smaller (to achieve finer sampling and more accurate curves), (6.8) shows that the difference between $V(i)$ and $V(i+1)$ also gets smaller. In fact, as Δt approaches zero, $V(i+1)$ approaches $V(i)$. Therefore, the two samples have essentially the same value at that point. This leads to the following conclusion: *The curve for volume versus time must be continuous; there can be no discontinuous jumps at any time.* The only way there could be a jump in V is if the flow rate Q was infinite, which is impossible in practice. The lower line in Fig. 6.2 shows the continuous nature of the volume V. (It should be noted, however, that the flow rate Q itself can have discontinuous jumps without violating any rule.)

The situation for pressure P is similar but a little more complicated when applied to a compliant vessel. The relationship between pressure and flow in a compliant vessel was found in Section 4.5 to be

$$\frac{dP}{dt} = \frac{1}{C}Q \qquad (6.9)$$

(This equation assumes the compliance C is constant.) In Euler's finite-difference form, (6.9) becomes

---

[1]There are several variations of the finite-difference technique, such as the predictor-corrector method and the Runge-Kutta method, which can be more efficient and accurate than Euler's method but which are somewhat more complicated. We will leave these methods to later courses.

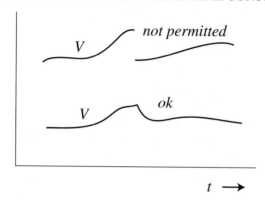

Figure 6.2: The volume V cannot jump instantaneously; its curve must be continuous.

$$P(i+1) = P(i) + \frac{\Delta t \cdot Q(i)}{C}. \tag{6.10}$$

The previous arguments can again be applied to show that (unless Q is infinite, which is impossible): *The curve for a vessel's pressure P versus time must be continuous; there are no discontinuous jumps.* This behavior is similar to the lower line in Fig. 6.2.

For most vessels, the compliance C is constant. If the vessel's compliance changes, as it does in active vessels such as the heart ventricles, an additional term is needed in (6.9), but because the compliance will change in a continuous manner, it can be shown that the pressure waveform must still be continuous.

6.4 FIRST-ORDER TIME CONSTANTS

Fluid systems that contain both resistive and compliant components (such as windkessel elements) are called first-order systems because the differential equations that describe any of their variables (pressure, volume or flow) have only first-order derivatives in them[2]. For example, a very simple fluid system composed of only one compliant element and one resistance element is shown in Fig. 6.3, using symbols.

To the right of the resistive element R in Fig. 6.3 is the symbol for a switch that closes at time $t = 0$. In fluid systems, this is usually a valve that blocks flow before $t = 0$, then is changed to let flow freely through at $t = 0$. The purpose of the valve (switch) is merely to start the flow—and the clock—at $t = 0$. Before the valve is thrown, there is some initial volume of fluid (it could be zero) in the vessel. Since the flow can't go anywhere before $t = 0$, this initial volume is trapped until the valve is thrown. This volume will result in some initial pressure inside the vessel (it could be zero)

---

[2]If inertial effects are included by involving the mass of the moving fluid, the differential equations are second-order, and the system is then called a second-order system. These systems can exhibit oscillatory behavior.

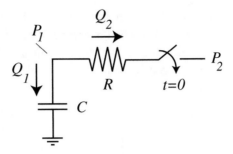

Figure 6.3: Simple first-order fluid system.

before and just at the time $t = 0$. Call this initial constant pressure P_i. After the valve is thrown, P_1 can then change. In the example of Fig. 6.3, the pressure P_2 is constant.

Let's write some equations relating the flow variables to each other at times *after* $t = 0$: From Poiseuille's principle,

$$Q_2 = (P_1 - P_2) / R. \tag{6.11}$$

From conservation of volume,

$$Q_1 = -Q_2. \tag{6.12}$$

From compliance relations [see (6.9)],

$$\frac{dP_1}{dt} = \frac{Q_1}{C}. \tag{6.13}$$

Substituting (6.11) into (6.12) then putting this into (6.13) results in the first-order differential equation for the system's pressure:

$$\frac{dP_1}{dt} = \frac{P_2}{RC} - \frac{P_1}{RC}. \tag{6.14}$$

Remember that P_1 is a variable while P_2 is a constant. A well-known *analytic* solution to (6.14) which meets the initial condition that $P_1 = P_i$ at $t = 0$ is found from calculus to be:

$$\boxed{P_1 = (P_i - P_2)e^{-\left(\frac{t}{\tau}\right)} + P_2.} \tag{6.15}$$

The constant τ in the denominator of the exponent in this solution is called the **time constant** of the system. The name comes from its role in determining how fast the flow variables in this system can change from one value to another. τ has the dimension of time.

Figure 6.4 is a plot of the pressure waveform found from (6.15). Notice that, as required, the pressure has an initial value of P_i before $t = 0$. Then when the valve is thrown at $t = 0$, the pressure P_1 starts changing in an exponential fashion on its way to the final value of P_2. (As stated earlier, since C is constant, the pressure curve is continuous in this example with no discontinuous jumps.)

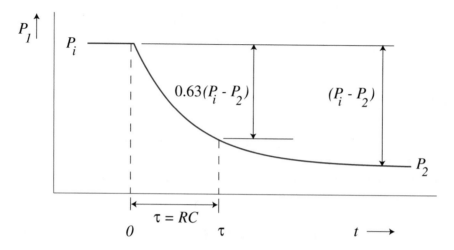

Figure 6.4: The pressure waveform in a first-order fluid system changes in an exponential manner. It reaches 63% of its excursion in one time constant τ.

The rate at which the pressure changes can be measured by the time it takes to get to a certain percentage, say 63%, of its total excursion[3] toward its final value P_2. In the example of this section, the pressure excursion is $(P_i - P_2)$, and P_1 will cover $(1 - e^{-1}) = 0.63 = 63\%$ of this excursion in a time $t = \tau$, as shown in Fig. 6.4. Thus, the larger τ is, the slower P_1 will change; the smaller τ is, the faster P_1 will change.

The value of τ for this system can be found by substituting (6.15) back into the differential Equation (6.14) and solving for τ. The result is

$$\boxed{\tau = RC} \qquad \text{Time Constant} \qquad (6.16)$$

This is a very important result: *The time response of a fluid system is given by the product of its compliance and its resistance* (or equivalent resistance if there are several resistive elements in series or parallel with the compliance). The larger the resistance and/or the compliance, the slower the

---

[3] P_1 actually never mathematically reaches P_2, even for an infinite time. It only approaches P_2 asymptotically, getting closer and closer as time progresses. Practically, though, it gets as close as you would want by just waiting a long enough time. In five time constants ($t = 5\tau$), it has gone 99.3% of the way to its final value, so 5τ is often used as a practical value for specifying the total length of time for the change.

fluid system will respond to changes and the less faithfully it will follow rapid variations in pressure, volume or flow. This is why the pulsatile pressure waveform of the blood entering the aorta is damped by the time it reaches the high-compliance venous side of the circulation, becoming essentially steady with time.

Although the results in the example above were obtained specifically for the pressure P_1, other fluid quantities in the system (the volume and flow rate) also exhibit exponential behavior, with the same time constant $\tau = RC$. Also, more complex first-order systems which have more than one compliant element will have similar exponential responses for all their fluid variables, but generally with several time constants, one for each compliant element.

Example 6.1.   Pressure Waveform in a Vein
A major leg vein has a valve at one end that allows blood to flow from one end toward the heart, but blocks flow in the opposite direction in Fig. 6.5. The pressure P_3 is constant at 10 mmHg, but the pressure P_1 changes in the manner (a simplifying approximation) shown in Fig. 6.6. The pressure P_2 is measured at the middle of the vessel's length. At time $t = 0$, P_2 is 10 mmHg and the volume inside the vessel is 60 mL. The total resistance R_t of the vein is 800 mmHg·s/L. The residual volume of this vessel is 10 mL.

Tasks

a. Write equations relating the problem's variables to each other.

b. Put these equations into sampled or vectorized form, using Euler's method to solve for the updated volume in the vessel, V_2.

c. Using Matlab, find and plot the waveforms of the volume V_2 and the pressure P_2 from $t = 0$ to 5 seconds. Also plot P_1 on the same graph as P_2. Calculate and display on this graph the total flow (total *volume* of flow) through the valve in those 5 seconds.

d. From the plot of P_2, estimate the vein's time constant.

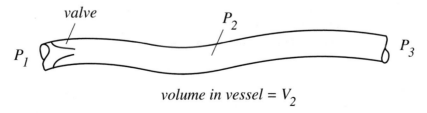

volume in vessel = V_2

Figure 6.5: Sketch of vessel analyzed in Example 6.1.

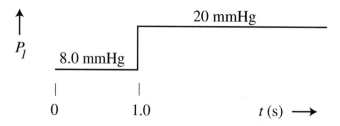

Figure 6.6: Specified variation of pressure P_1.

Solution
We'll model this vessel as a simple windkessel element with a valve, as shown below in Fig. 6.7.

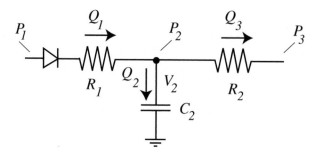

Figure 6.7: Schematic model of vessel.

First find the value of compliance C_2. From Equation (4.7) in the Hooke's Law chapter applied at $t = 0$,

$$C_2 = (V_2 - V_\phi)/P_2 = (0.050 \text{ L})/(10 \text{ mmHg}) = 0.0050 \text{ L/mmHg}.$$

Assume that $R_1 = R_2$ since P_2 is near the middle of the vessel. Since these resistance elements are in series, $R_t = R_1 + R_2 = 2R_1 = 800 \text{ mmHg} \cdot \text{s/L}$. Thus, $R_1 = R_2 = 400 \text{ mmHg} \cdot \text{s/L}$. The pressure P_3 will be constant, but P_1 jumps from 8.0 mmHg to 20 mmHg at $t = 1$s.
a. Now write equations relating the variables of the problem to each other –
The pressure P_2 due to volume V_2 in compliant vessel:

$$P_2 = (V_2 - V_\phi)/C_2. \tag{6.17}$$

The flow rate through R_1, including valve action:

$$\text{If } P_1 > P_2, \qquad Q_1 = (P_1 - P_2)/R_1. \text{ Otherwise } Q_1 = 0. \tag{6.18}$$

Flow rate through R_2:

$$Q_3 = (P_2 - P_3)/R_2. \tag{6.19}$$

Flow rate into the compliant element using conservation of volume:

$$Q_2 = Q_1 - Q_3. \tag{6.20}$$

The change in volume V_2 due to flow into the vessel:

$$\Delta V_2/\Delta t = Q_2. \tag{6.21}$$

b. Put these equations into sampled or vectorized form –
Equation (6.17) will become:

$$P_2(i) = (V_2(i) - V_\phi)/C_2.$$

Equation (6.18) becomes:

$$\text{If } P_1(i) > P_2(i), \qquad Q_1(i) = (P_1(i) - P_2(i))/R_1. \quad \text{Otherwise } Q_1(i) = 0.$$

Equation (6.19) becomes:

$$Q_3(i) = (P_2(i) - P_3)/R_2.$$

Equation (6.20) becomes:

$$Q_2(i) = Q_1(i) - Q_3(i).$$

Equation (6.21) can be reformulated using Euler's method:

$$V_2(i + 1) = V_2(i) + Q_2(i)\Delta t.$$

[Notice that we have vectorized *all* of the variable flow rates (Q_1, Q_2, and Q_3) and the two variable pressures (P_1 and P_2), but *not* the constant pressure P_3 or residual volume V_ϕ.]

c. We can now use a Matlab script m-file (label it euler.m) to calculate the waveform of the volume V_2, the pressure P_2, and find the total flow V_t. Since the total time interval we are interested in is 5 seconds, let's break it into 500 increments, each of $\Delta t = 0.01$ s. Also, let's store the problem's constants in a separate m-file, named paramex.m. The paramex.m file will look like this:

```
% Parameter file for the example in Euler's Method chapter
R1=400; R2=400;   % units: mmHg-s/L
C2=0.005;         % units: L/mmHg
V0=0.010;         % residual vol, in L
P1s=8;   P1d=20;  % start and end values of P1, in mmHg
```

```
dt=0.01; N=500;      % time increment, 5 seconds total
P3=10;               % constant pressure at outlet, in mmHg
V2(1)=0.060;         % initial value of volume at t=0, in L
```

The script m-file called euler.m will look something like this:

```
%  Example of Euler's method - Program for finding and plotting the
%  volume and pressure in a compliant and resistive vessel connected
%  to a source, and finding the total volume flow in 5 seconds.

input('What parameter file do you want to use?'); % type paramex, then hit return

%  Set up step-function vector for input pressure:
P1(1:100) = P1s;
P1(101:500) = P1d;
%  Step through time for N increments:
for i = 1:N;

      %  Find pressure from volume using compliance:
      P2(i) = (V2(i) - V0) / C2;

      %  Use Poiseuille's law to calculate flows:
      if P1(i) > P2(i); Q1(i) = (P1(i)-P2(i)) / R1;    % flow through R1
              else Q1(i) = 0;
      end      % end of if
      Q3(i) = (P2(i)-P3) / R2;      % flow through R2

      %  Apply conservation of volume:
      Q2(i) = Q1(i) - Q3(i);

      %  Use Euler's method to update volume:
      V2(i+1) = V2(i) + Q2(i)*dt;
      %  Note-because of Euler's forward method, V2 has 501 elements

end    %  loop back until i=N

%   Plot volume:
tv = [0:dt:N*dt]; % set up time vector for plotting volume (501 values)
hold off;  plot(tv,V2);
xlabel( 'time (s)' );  ylabel( 'V2 (L)' );

%   Now plot two graphs for pressure on the same (new) figure:
t = [0:dt:(N-1)*dt];   % set up time vector for plotting pressures (500 values)
figure; plot(t,P1,'-b');  hold on;  plot(t,P2,'--r');
xlabel( 'time (s)' );  ylabel( 'P1 and P2 (mmHg)' );

%   Now calculate total flow volume:
Vt = sum(Q1)*dt;        % total volume of flow through valve in 5 sec
Vstr = num2str(Vt,2);   % convert number to a string with 2 sig figs
```

```
text(3,18,[ 'Total flow = ' Vstr ' L'])          % place string inside plot
%   (Note necessary spaces on either side of Vstr in previous line)
text(1.2,14,[ 'P1' ]); text(4.1,14,['P2' ]);   % put labels near each line
```

The plots will look like those in Fig. 6.8.

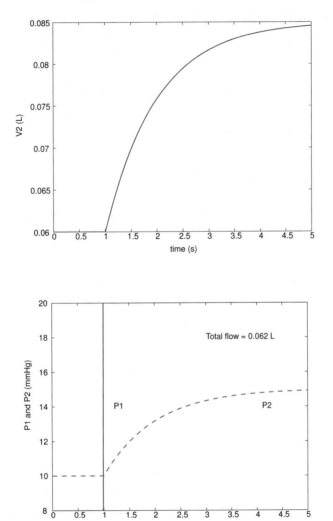

Figure 6.8: Plots of volume and pressures in Example 6.1.

d. In the pressure plot above P_2 changed in an exponential manner from 10.0 to 15.0 mmHg, for a total excursion of $15.0 - 10.0 = 5.0$ mmHg. Now 63% of this excursion will be at a pressure of

$10.0 + 0.63(5.0) = 13.2$ mmHg. This occurs at the time $t = 2.0$ s, which is 1.0 s after P_1 jumped. So the time constant is $\tau = 1.0$ s.

6.5   PROBLEMS

6.1. m. The vena cava returns blood to the right side of the heart through the tricuspid valve. This valve allows blood to go out of the vena cava (i.e., out of end 3) but blocks flow in the reverse direction. A simple diagram is shown in Fig. 6.9.

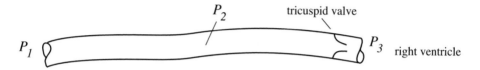

Figure 6.9: Sketch of vessel analyzed in Problem 6.1.

The pressure P_1 is constant at 11 mmHg, but the pressure P_3 (the pressure in the right ventricle) changes. Over a diastolic/systolic cycle, the pressure P_3 has the following approximate form as in Fig. 6.10.

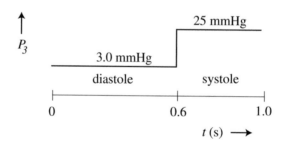

Figure 6.10: Specified variation of pressure P_3.

The pressure P_2 is measured near the middle of the vessel's length. At time $t = 0$, the total blood volume inside the vena cava is 475 mL. Let the residual volume be 60 mL. The compliance of the vena cava for this problem is $C_v = 0.050$ L/mmHg. The total resistance of this vessel is 60 mmHg·s/L. (Note: These values are valid for this problem only; they will be different for the Major Project.)

Tasks:

a. Draw a *windkessel* model of this vessel (similar to the one on p. 80).

b. Write equations relating the variables of the problem (all Q's, P's and V_2) to each other.

c. Put these equations in sampled or vectorized form, ready for Matlab coding, using Euler's method to update the volume V_2.

d. Use Matlab to find and plot the waveform for the pressure P_2, from $t = 0$ to 1.0 second.

e. Calculate the average volumetric flow rate in units of L/min (not the total volume) through the tricuspid valve during this 1.0 s period. Print out this value inside the plot.

[All you need to turn in for this problem are printouts of your Matlab m-files, and your plot, shown below]

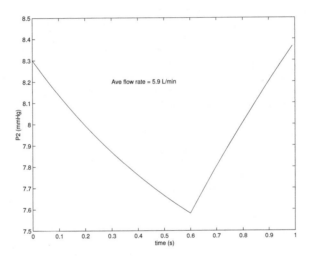

Figure 6.11: Plot of Matlab calculation of pressure P_2.

CHAPTER 7

Muscle, Leverage, Work, Energy and Power

7.1 INTRODUCTION – MUSCLE

The human body uses movement in numerous ways, such as walking, finding food, moving food through the intestines, breathing, looking around, pumping blood through the circulation, and in countless other functions. Almost all of the motion is achieved by muscle. Muscle tissue is found throughout the body (as you might suspect from the list above), such as the attachment of striated muscle to bones for lifting, walking, chewing and breathing; the smooth muscle in the walls of the blood vessels (especially the arterioles) for regulating blood pressure and directing flow to vital organs; and, of course, the cardiac muscle of the heart for providing driving pressure in the ventricles for circulating blood throughout the body.

The basic mechanism for muscle contraction can be seen by examining the molecular structure of a submicroscopic unit of muscle, the sarcomere, shown in Fig. 7.1. The sarcomere unit, about 2.2 μm in length, consists of long molecules that interdigitate: the thin filaments, made of actin, which overlap with the thick filaments, made of myosin. When the muscle is innervated to contract, calcium ions flood the region, causing the thick and thin filaments to attract each other and pull themselves into a condition of greater overlap. This forms the "sliding filament" basis of muscle contraction. Since a very large number of sarcomere units act together over the length and cross-section of the muscle bundle, a large force is produced along the axis of the muscle. If allowed, the muscle will shorten and movement will result.

By its molecular nature, a muscle is capable of producing large forces, but the overall extent of its shortening, in absolute distance, is not very great. This characteristic of muscle makes it well suited for many of its functions in the body—such as ventricular contraction—but acting alone it is not well matched to the needs of the skeleton for walking and lifting. Many of these actions require larger distances of movement than the muscle alone can provide, but with less force. **Levers** provide a way of converting small movement at one point on the lever to larger movement at other points. We will see next how muscles use levers to advantage in the musculoskeletal system—the interconnected muscles and bones of the body.

7.2 LEVERS AND MOMENTS

Levers are rigid shafts that pivot around some point, called the **fulcrum**. Forces are applied at various points on the lever, and if unbalanced, can cause the lever to rotate around its fulcrum. We

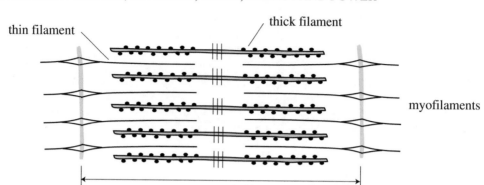

Figure 7.1: A sarcomere unit of muscle, showing overlap of the thick and thin filaments.

are interested here in situations where the forces are opposing and balanced so the lever is stationary, or almost stationary, such as when the body's skeleton is being held in a static posture.

The forces on the lever usually consist of a load or weight W and an applied pulling force F. There are three possible configurations for the relative locations of these forces in relation to the fulcrum; these are classified as Class 1, Class 2, or Class 3 levers, summarized in Fig. 7.2. The Class 1 levers are characterized by having the forces W and F on opposite sides of the fulcrum, but in the same direction. Class 2 levers have the forces on the same side of the fulcrum, but in opposing directions, with the pull F acting at a point further from the fulcrum than the load W. Class 3 levers also have the forces on the same side of the fulcrum, but the load W is further from the fulcrum than the pull F. Using bones as the levers, the human body has examples of Class 1 and Class 3 levers at different locations, but most skeletal levers are of type Class 3.

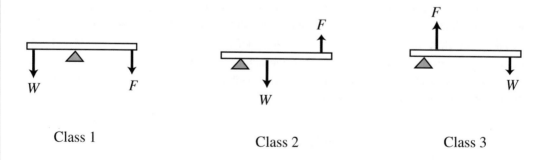

Figure 7.2: The three classifications of levers. Most musculoskeletal levers are of Class 3.

The capability of each force to produce rotation of the lever is not directly equal to the force, but rather to the **moment** M of the force, which is the product of the component F_n of the force normal (i.e., perpendicular) to the lever's axis and the distance d between the fulcrum and the point of action of the force. The quantity d is often called the "lever arm." Thus,

$$M = F_n d.$$
Moment
(7.1)

We will use the sign convention that if a force component F_n is attempting to rotate the lever around the fulcrum in a *clockwise* direction, the sign of its moment is positive. If it is attempting to rotate the lever in a *counterclockwise* direction, the sign of its moment is negative.

To balance a lever and keep it stationary, the moments of all the forces on the lever must balance; that is, their sum must be zero:

$$\sum_{i=1}^{N} M_i = \sum_{i=1}^{N} F_{ni} d_i = 0$$
In balance
(7.2)

As an example, Fig. 7.3 shows the case of a simple lever with two forces, a lifting force F and a weight W, on the same side of the fulcrum but in opposite directions—a Class 3 lever. It is representative of the forearm lifting an object held in the hand with the biceps muscles providing the lifting force. The biceps muscles attach to the forearm bone through a tendon at a point that is closer to the fulcrum with the upper arm bone (at the elbow) than is the weight in the hand. Due to the differences in the lever arm lengths, the muscle force F must be greater than the weight W.

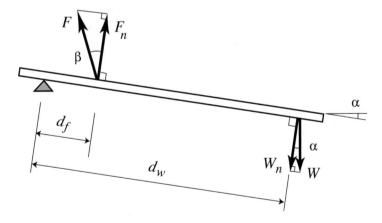

Figure 7.3: A model of a lever representing the forearm at an angle, with the force produced by the biceps muscles (F) opposing a weight (W) held in the hand.

Figure 7.3 shows the case where the lever is not exactly horizontal; the lever axis forms an angle α with respect to the horizontal. This means that the force of the weight W, directed vertically downward by the pull of gravity, is not perfectly normal to the axis of the lever. In addition, the pulling force F is at an angle β with respect to the lever's axis. We must first find the components of each of these forces in the direction normal to the axis of the lever. By trigonometry,

$$F_n = F \cos \beta \tag{7.3}$$

and

$$W_n = W \cos \alpha. \tag{7.4}$$

Then the moment due to the weight is given by $M_w = W_n d_w = W d_w \cos \alpha$, and the moment of the lifting force is $M_f = F_n d_f = F d_f \cos \beta$. Using (7.2) to balance the moments,

$$- F \, d_f \cos \beta + W d_w \cos \alpha = 0. \tag{7.5}$$

Solving for F,

$$F = W(d_w \cos \alpha)/(d_f \cos \beta). \tag{7.6}$$

Since α and β are small and often close to each other in value, and since $d_w \gg d_f$, then $F \gg W$. A numerical example of this relationship for a forearm of the body is given next.

Example 7.1.   Leverage of Forearm

A book weighing 5.0 pounds is being held steadily in the hand at a slight downward angle as shown in the figure below. Using the typical values given for the length of the forearm and the insertion point of the biceps muscles onto the forearm, calculate the force F that the biceps muscles must produce to hold the book steady. See Fig. 7.4.

Solution

The model for this configuration is the same as shown in Fig. 7.3. The equation for the force F is given by (7.6):

$$F = W(d_w \cos \alpha/(d_f \cos \beta) = 5.0 \, (14 \cos 15°)/(2.0 \cos 30°) = \textbf{39 lbf}.$$

Therefore, it takes a biceps-muscle force of 39 pounds to hold a 5-pound book. Converting this force to SI units,

$$F = (39 \text{ lbf}) \left(\frac{4.448 \text{ N}}{1 \text{ lbf}} \right) = \textbf{170 N}.$$

As just shown, because the lever arm of the weight is much longer than the lever arm of the force in this Class 3 lever, $F \gg W$. But if the lever is now allowed to rotate, the *distance* traveled by

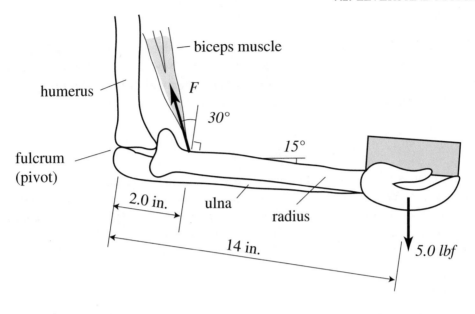

Figure 7.4: Sketch of lever arm analyzed in Example 7.1.

the weight will be much *greater* than the distance traveled by the point of the lever where the force is attached, by the ratio d_w/d_f. This is shown in Fig. 7.5. This is how the relatively small contractions of muscle can be converted to much larger motions at the end of bones acting as Class 3 levers.

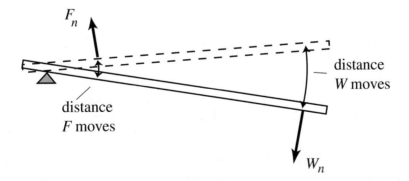

Figure 7.5: In a Class 3 lever, the load W moves further than the pulling force F. Since muscle cannot contract a long distance, this is a way that muscle can affect a longer movement, but more force is needed.

7.3   WORK

"Work" is a word familiar to us all ("That was hard work..."). In fact, the quantitative definition of work is consistent with our everyday experience, that of using force to accomplish movement over some distance, and of expending an amount of energy in doing so. Mathematically, the quantity of work is equal to the force F applied to a body times the distance d the body is moved in the direction of the force, or

$$\text{Work} = Fd, \tag{7.7}$$

where F is the component of force in the same direction as the movement d.

The units of work are therefore equal to the units of force (N in SI units) times the units of distance (m), or N·m. The units N·m are called joules, with the symbol J. These are the same as the units of energy, and since the units are the same, you might expect that there is a relationship between work and energy. We will return to this point after a discussion of energy.

7.4   ENERGY

Energy is a quantity that represents the overall excitation state of a system. The **First Law of Thermodynamics** states that energy can neither be created nor destroyed overall; that is, in any closed system (no energy crossing its borders), *energy must be conserved*. It can be changed from one form to another, but the total energy of the system must remain constant. For purposes of this class, we will define three forms of energy:

1. **Kinetic Energy** – If a body of mass m is moving with velocity v, its kinetic energy is given by a single formula:

$$E_k = 1/2 \, mv^2. \tag{7.8}$$

In calculating the kinetic energy, the direction of the velocity does not matter (since it is squared), just its magnitude.

2. **Potential Energy** – Since potential energy represents the energy stored in various components of a system and therefore can be found in various formats, there is not a single equation for potential energy that covers all cases. Each type of potential energy needs to be treated individually. We will look at only three cases:

   a. Springs – When a spring is wound (or extended or compressed), the energy used to wind it is stored as potential energy. It can be converted back to other forms of energy at a later time. There are numerous examples of springs or spring-like mechanisms in nature and in the body, including the compliance behavior of blood vessels and the action of elastic tissues (ligaments, bladders, and such) in the body.

   b. Chemical – The energy that is released from chemical species upon reaction with other chemicals is a type of potential energy. An extremely important example of this for living organisms is the energy available from food when it is metabolized. This food energy is transformed into kinetic energy (movement), other potential energy (ATP molecules, stretched elastic tissue), or heat in the body.

c. Gravity – When a mass is moved higher against the force of gravity, its potential energy E_p increases by an amount

$$\Delta E_p = mgh, \tag{7.9}$$

where

m is the body's mass,

g is the acceleration of gravity (9.8067 m/s^2), and

h is the height that the mass is raised, measured in a vertical direction parallel to g.

An example of this is shown in Fig. 7.6.

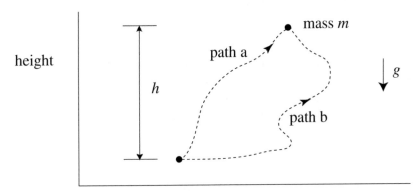

Figure 7.6: Potential energy of mass m is increased when the mass m is moved a height h against the force of gravity.

Notice that the change in potential energy depends only on the height h and not on the particular path taken in reaching that height (i.e., it is the same for both path a and path b in Fig. 7.6. Thus, gravity is called a **conservative** field. (If the quantity being calculated depends on the particular path taken, such as the energy needed to overcome friction, the force is called non-conservative.)

3. **Heat** – This is the third form of energy we'll consider, although strictly speaking, heat is a manifestation of increased vibrational energy (and thus is really kinetic energy) of the molecules making up a system. Heat is sometimes a useful form of energy (as with steam engines or increased chemical reaction rates and diffusion), but often it is wasted to the outside environment. In many

systems, the other two forms of energy (kinetic and potential) are easily transformed into heat, but it is more difficult to extract useful energy out of heat. This is why we classify it separately.

Returning now to the concept of work, how does work (which has the units of energy) relate to the above forms of energy? Work is a convenient way of *measuring* the transfer of energy from one form to another. For example, Fig. 7.6 can be used to derive (7.9) by using the measure of work. The force needed to raise the mass m against gravity is $F = mg$, in the opposite direction to g. Since *work = force × distance*, the work needed to raise the mass m to a height h is just mgh, exactly equal to the increase in potential energy given by (7.9).

7.5   POWER

When a system transfers an amount of energy ΔE during a time Δt, the *rate* of energy exchange is given by $\Delta E / \Delta t$. This is the power Φ being transferred, or

$$\Phi = \Delta E / \Delta t. \qquad \text{Power} \qquad (7.10)$$

In the limit as Δt approaches zero, we get the differential form of (7.10):

$$\Phi = dE/dt. \qquad \text{Power} \qquad (7.11)$$

From either (7.10) or (7.11), the units of power can be seen to be J/s, or W (watts). This is a unit we can relate to everyday experience, since incandescent light bulbs for room lighting are usually rated from 40-100 W (electrical input power), while flashlight bulbs are 1-10 W. Also, since 1 hp (horsepower) is equivalent to 746 W, a 100 hp automobile engine can put out 74,600 W of power at maximum output.

7.5.1   POWER IN FLUID FLOW

As covered in Chapter 3, it takes energy to force fluid through a tube against its resistance, and since power is defined as the energy expended per unit time, there is a power requirement for maintaining the flow. The relationship between power and the fluid parameters can be seen by considering the following:

```
power = energy/time = work expended/time = force X distance/time =
   pressure drop X area X distance/time = pressure drop X volume/time =
   pressure drop X volume flow rate.
```

Thus, the power Φ is given by the product of the pressure drop and the volume flow rate:

$$\Phi = \Delta P \cdot Q. \qquad (7.12)$$

The units on the right side of (7.12) are $(N/m^2)(m^3/s) = N \cdot m/s = J/s = W$, which of course are the same as the units on the left side of (7.12). Relating the pressure drop and flow rate to resistance by (3.7) in the Poiseuille's Law unit gives alternate forms for the expended power:

$$\Phi = Q^2 R = \frac{(\Delta P)^2}{R}.$$ (7.13)

7.6 PROBLEMS

7.1. In the country of Combria, blood pressure is measured in units of Pa. Knowing that the density of mercury is 13.6 g/mL, find the conversion factor from Pa to mmHg. (Do not obtain the conversion factor from a table; rather, calculate it.)

[ans: 1 Pa = 7.50×10^{-3} mmHg]

7.2. In the Example 7.1 on p. 90, the forearm (radius) is held at an angle of 15° from the horizontal and the biceps muscle forms a 30° angle with the forearm (measured to the normal). Suppose that the arm holding the book is now extended forward from the body (but still held at 15° from horizontal), causing the angle of the biceps muscle with respect to the forearm to change to 80.3° (measured to the normal). What force must the muscle now apply to hold the book steady?

[ans: 890 N]

7.3. The masseter muscle is the major muscle group attached to the mandible (jaw bone) for chewing food, as shown below:

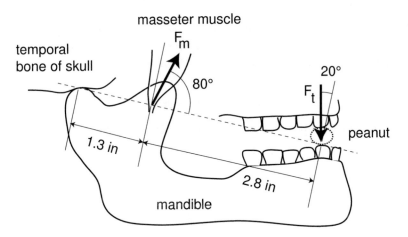

Figure 7.7: Sketch of lever analyzed in Problem 7.3.

It takes a force of about 5.5 lbf to crack open a peanut with the teeth. How much force does the masseter muscle need in order to crack open the peanut using the typical values given in the figure above?

[Hint: First draw a static lever (moment) diagram like Fig. 7.3. ans: $F_m = 74$ N]

7.4. You take a bicycle trip from the mouth of Little Cottonwood Canyon in Utah (starting at an elevation of about 5200 feet) up to the lodge at Snowbird resort. The trip takes 2.0 hours.

a. How much mechanical energy has your body provided to increase your potential energy during the trip? (Ignore frictional loses in the bicycle—not a very valid assumption. Also look up the elevation of the Snowbird lodge on their website.)

[ans for me: 6.1×10^5 J]

b. Your body burns food (chemical energy) as a source of its mechanical energy. The maximum efficiency of this process is about 20%; that is, it takes 5 J of chemical energy to produce 1 J of mechanical energy output from the body at a maximum work rate. Considering this, how many bags of peanut M&M's (1.74 oz. bags) would you need to eat to provide the energy for climbing the hill?

[ans for me: about 3 bags]

c. Instead of bicycling to Snowbird, if you were to connect your bicycle to a stationary electrical generator and peddle at the same rate, could you light up a 100 W light bulb? (Assume the efficiency of the generator is 100% in converting mechanical energy to electrical energy.)

7.5. a. Calculate the average power expended by the left ventricle in pumping blood around the human CV system. (Use the average cardiac output and average pressure values for the aorta and the right atrium found in the Major Project figures.)

[ans for my values: 1.2 W]

b. Would this amount of power light up a typical hallway nightlight?

APPENDIX A

Conversion Factors

1 lbf = 4.448 N

1 psi = 6895 Pa

1 slug = 14.59 kg

1 dyn = 1×10^{-5} N

1 dyn/cm^2 = 1×10^{-1} Pa

1 atm (atmospheric pressure at sea level) = 1.013×10^5 Pa = 1.013 bars
$$= 14.70 \text{ psi} = 760.0 \text{ mmHg} = 29.92 \text{ inHg}$$

1 mmHg at 0°C = 133.3 Pa = 1.000 torr

1 inH$_2$O at 4°C = 249.1 Pa

1 P (poise) = 1×10^{-1} Pa·s = 1×10^{-1} N·s/m^2 = 1×10^{-1} kg/(m·s)

1 Å (angstrom) = 1×10^{-10} m = 1×10^{-1} nm

1 L = 1.000×10^{-3} m^3

1 gal = 3.785 L

1 Da = 1.661×10^{-27} kg

1 in = 2.540 cm

1 m = 39.37 in = 3.281 ft

1 W = 1 J/s = 1 (N·m)/s

1 erg = 1×10^{-7} J

1 hp = 0.7457 kW

1 Btu = 1055 J

1 cal ("small" calorie) = 4.184 J

1 Cal ("big" calorie, food calorie, or kilocalorie) = 1×10^3 cal

Notes:

1. When the numeral "1" appears alone in the above formulas (e.g., 1×10^{-5} N, or 1 L), it is considered exact with an infinite number of significant figures.

2. The food and nutrition industry uses the big calorie, or Cal, as the unit of chemical energy available from food, but often Cal is not capitalized on food labels so don't confuse it with cal, the small calorie.

APPENDIX B

Material Constants

B.1 VISCOSITY

Dynamic viscosity μ has SI units of kg/(m·s). These units are equivalent to Pa·s or N·s/m^2. However, a non-SI unit is commonly used to report viscosity: the poise (P), named in honor of Poiseuille. 1 P is equal to 1×10^{-1} Pa·s. Approximate viscosities of some common fluids are listed below at standard temperature and pressure in units of either P or cP (centipoise).

Viscosity	
(1 P = 1 × 10^{-1} Pa·s)	
(1 cP = 1 × 10^{-2} P = 1 × 10^{-3} Pa·s)	
air	1.8×10^{-4} P
glycerine (20°C)	1500 cP
water (20°C)	1.0 cP
blood plasma (Newtonian)	1.2 cP
whole blood (non-Newtonian)	2.5 - 14 cP
hemoglobin	6 cP
egg protoplasm	1.8 cP

B.2 DENSITY AND SPECIFIC GRAVITY

Density ρ is mass per unit volume and has SI units of kg/m^3. Sometimes the density is reported as the *specific gravity* γ, which is defined as the density of the material divided by the density of water.

Density	
air at standard atm	1.22 kg/m^3
water at 15°C	999 kg/m^3
whole blood at 37°C	1060 kg/m^3

Specific Gravity	
red blood cells	1.10
blood plasma	1.03

B.3   PERMEABILITY

Darcy's permeability k has SI units of m^2; in this table it is given in units of cm^2. The permeabilities of some example materials are listed below.

Permeability	
sand	approx. 1.0×10^{-6} cm^2
cellulose filter paper (11 μm pore size)	1.2×10^{-9} cm^2
granite	approx. 1.0×10^{-16} cm^2
cell membrane	approx. 1.0×10^{-17} cm^2
skin	approx. 1.0×10^{-18} cm^2

B.4   YOUNG'S MODULUS AND ULTIMATE STRESS

Young's Modulus E (also known as elastic modulus or stiffness) has SI units of Pa; it is given below in GPa.

The Ultimate Stress is the maximum stress a material will sustain before breaking; it is given below in MPa.

Young's Modulus and Ultimate Stress for Various Orthopedic Materials		
Material	Ultimate Stress (MPa)	Young's Modulus E (GPa)
Stainless steel	850	180
Cobalt alloy	700	200
Titanium alloy	1250	110
PMMA	35	5
Polyethylene (high MW)	27	1
Bone (cortical)	120	18

Bibliography

[1] Berger, S.A., Goldsmith, W. and Lewis, E.R. 1996. *Introduction to Bioengineering*, Oxford: Oxford University Press.

[2] Berne, R.M. and Levy, M.N. 1993. *Physiology*, St. Louis: Mosby-Year Book, Inc.

[3] Darcy, H. 1856. *Les Fountaines de la Ville de Dijon*, Dalmont, Paris.

[4] Durney, C.H. 1973. "Principles of Design and Analysis of Learning Systems," *Engineering Education*, March 1973, 406-409.

[5] Durney, C.H. and Christensen, D.A. 2000. *Basic Introduction to Bioelectromagnetics*, Boca Raton: CRC Press.

[6] Eide, A.R., et al. 1997. *Engineering Fundamentals and Problem Solving*, 3rd Edition, Boston: WCB McGraw-Hill.

[7] Enderle, J., Blanchard, S. and Bronzino, J. 2000. *Introduction to Biomedical Engineering*, San Diego: Academic Press.

[8] Folkow, B. and Neil, E. 1971. *Circulation*, New York: Oxford University Press.

[9] Fung, Y.C. 1984. *Biomechanics: Circulation*, New York: Springer-Verlag.

[10] Fung, Y.C. 1990. *Biomechanics: Motion, Flow, Stress, and Growth*, New York: Springer-Verlag.

[11] Guyton, A.C. 1974. *Function of the Human Body*, Philadelphia: Saunders.

[12] Guyton, A.C. and Hall, J.E. 2000. *Textbook of Medical Physiology*, Philadelphia: Saunders.

[13] Keener, K. and Sneyd, J. 1998. *Mathematical Physiology*, New York: Springer.

[14] Mars Climate Orbiter, on NASA web site: www.NASA.gov.

[15] Matlab Help Desk, The Math Works Inc., Natick, MA.

[16] Nillson, J.W. and Riedel, S.A. 1996. *Electric Circuits*, 5th Edition, Reading, MA: Addison-Wesley.

[17] Palm, W.J. III. 1999. *Matlab for Engineering Applications*, Boston: WCB McGraw-Hill.

[18] Peterson, K. 1999. *A Numerical Simulation of the Cardiovascular System to Investigate Changes in Posture and Gravitational Acceleration*, MS Thesis, University of Utah.

[19] Senzaki, H., Chen, C-H. and Kass, D.A. 1996. "Single-Beat Estimation of End-Systolic Pressure-Volume Relation in Humans," *Circulation*, 94: 2497-2505.

[20] Silverthorn, D. 1998. *Human Physiology, an Integrated Approach*, Upper Saddle River: Prentice-Hall.

[21] Starling, E.H. 1918. *The Linacre Lecture on the Law of the Heart*, London: Longmans, Green.

[22] Withers, P.C. 1992. *Comparative Animal Physiology*, Fort Worth: Saunders College Publishing.